Laboratory Guide to Biochemistry, Enzymology, and Protein Physical Chemistry

A Study of Aspartate Transcarbamylase

Laboratory Guide to Biochemistry, Enzymology, and Protein Physical Chemistry

A Study of Aspartate Transcarbamylase

Marc le Maire

Centre National de la Recherche Scientifique
Gif-sur-Yvette, France

Raymond Chabaud

Université Pierre et Marie Curie
Paris, France

Guy Hervé

Centre National de la Recherche Scientifique
Gif-sur-Yvette, France

SPRINGER SCIENCE+BUSINESS MEDIA, LLC

This book is a translation, revised and enlarged by the authors, of *Biochimie—un modèle d'étude: l'aspartate transcarbamylase—théorie et guide d'expériences,* published by Masson in Paris in 1990.

ISBN 978-0-306-43639-0 ISBN 978-1-4615-3820-2 (eBook)
DOI 10.1007/978-1-4615-3820-2

Originally published by Plenum Press New York in 1991

Foreword

The study of a single well-chosen substance, here aspartate transcarbamylase, can provide an excellent basis for a laboratory course. The student is introduced to a variety of scientific ideas and to many experimental and interpretive techniques. This enzyme is readily available, is relatively stable, has an extensive literature, and its behavior has many facets: substrate inhibition, a large change in structure upon homotropic activation by substrates, allosteric stimulation by ATP, allosteric inhibition by CTP synergistic with UTP, positive cooperativity for substrates, negative cooperativity for CTP binding, and dissociation and reassembly of subunits C_3 and R_2 from the holoenzyme C_6R_6. In addition to the known biochemical aspects of these properties, the results obtained here can be interpreted in the light of the high-resolution X-ray diffraction structures of the *T* and *R* forms, the low-angle X-ray scattering results, and the large number of mutants now available by recombinant DNA methods. Future development of this course could also involve part of these methods, as well as the carefully chosen experiments described here.

This approach resembles research more than the approaches one usually finds in biochemical laboratory courses. A consistent development of ideas about a single enzyme, which shows so many facets in its behavior, is sure to hold the interest of the student. Moreover, one explores a depth, and reasons to move forward, that are an essential part of research.

Some years ago, I conceived and put into practice a somewhat similar plan for the laboratory part of a first semester of general chemistry. The student prepared a transition metal complex, made a chemical analysis, analyzed spectra (UV, visible, infrared, NMR), and then measured kinetics of a subsequent transformation of the complex. This program was very successful in that it incorporated preparative, analytical, and physical–chemical methods as essential parts of a larger purpose. I expect similar success with the plan so well described in this guide.

William N. Lipscomb

Harvard University

Preface

This book presents an integrated set of laboratory experiments devised for the teaching of biochemistry, enzymology, and protein physical chemistry. The program relies entirely on the use of a unique enzyme, aspartate transcarbamylase, which exhibits all of the catalytic and regulatory properties characteristic of allosteric enzymes. It can be easily and reversibly dissociated into subunits that have only the catalytic properties and thus behave as a simple Michaelian enzyme. A comprehensive study of this system leads to the use of numerous biochemical techniques.

Biochemistry and molecular biology are young sciences, under rapid development. It is desirable that students of these subjects have the benefit of concrete training directly connected to today's practices as developed in research laboratories. Aspartate transcarbamylase is presently the object of high-level fundamental research, and the following teaching program benefits directly from the results obtained. An up-to-date list of references guides the reader to more detailed information.

The book begins with a general presentation of aspartate transcarbamylase and of the concepts of cooperativity and allostery. A study of the expression of the gene coding for the enzyme is then described, followed by a chapter dealing with the purification of the enzyme and its dissociation into subunits. The next chapter describes a wide experimental program concerning the structure and physical chemistry of the enzyme and its subunits. Finally, an extensive study of the catalytic and regulatory properties of these proteins is proposed. Each chapter begins with a presentation of the underlying theory. Appendixes provide useful information, including constants and units. Within each experimental section several questions are posed, and answers are given at the end of the book.

Initially devised for the course Biochemistry and Cellular Biology and Physiology at the Pierre et Marie Curie University of Paris (Paris VI), this successful program is now used in other universities both in France and abroad. It is addressed primarily to undergraduate biochemistry students as well as to those following courses that involve biochemistry (medical schools; schools of pharmacy and agronomy). The book is sufficiently detailed to be of interest also to graduate students. It is a flexible program that can be adapted in terms of the number and nature of experiments performed in order to organize 50 to 200 hours of laboratory classes.

Teachers will find at the end of the book a nonexhaustive list of complementary pedagogic possibilities that are offered by the system studied but are not presented in detail in this edition. The authors are at your disposal should you wish to examine closely any of these possibilities, as well as those emerging from current research. Criticisms and suggestions from our teaching colleagues or students are welcome.

We wish to thank all of our friends who have already helped us with their advice and encouragement, in particular Dr. Anthony Else, who efficiently improved the English version of this book. We especially thank Ms. Jocelyne Mauger (Laboratoire d'Enzymologie of CNRS) for her skillful collaboration in its framing.

Marc le Maire
Raymond Chabaud
Guy Hervé

Gif-sur-Yvette and Paris

Contents

Chapter 6

COMPLEMENTARY EXPERIMENTS

APPENDIX

Chapter 1

Aspartate Transcarbamylase

Aspartate transcarbamylase (ATCase) catalyzes the carbamylation of the amino group of aspartate by carbamylphosphate according to the following reaction:

$$\text{CARBAMYL PHOSPHATE} + \text{ASPARTATE} \xrightleftharpoons{\text{ATCase}} \text{CARBAMYL ASPARTATE} + \text{PHOSPHATE}$$

This reaction is the first step of the pyrimidine pathway. This important enzyme is present in all organisms, and it is the ATCase from *Escherichia coli* that will be considered here. In this microorganism, the catalytic activity of ATCase is feedback inhibited by the end product CTP, which in this way limits its own production. On the other hand, its activity is stimulated by a purine, ATP. This antagonism is one of the phenomena which, in the cell, tend to balance the production of purines and pyrimidines for the biosynthesis of nucleic acids (Fig. 1).

As in most cases of feedback inhibition or activation, the effectors CTP and ATP have molecular structures that are very different from those of the substrates, and they bind to the enzyme on specific sites that

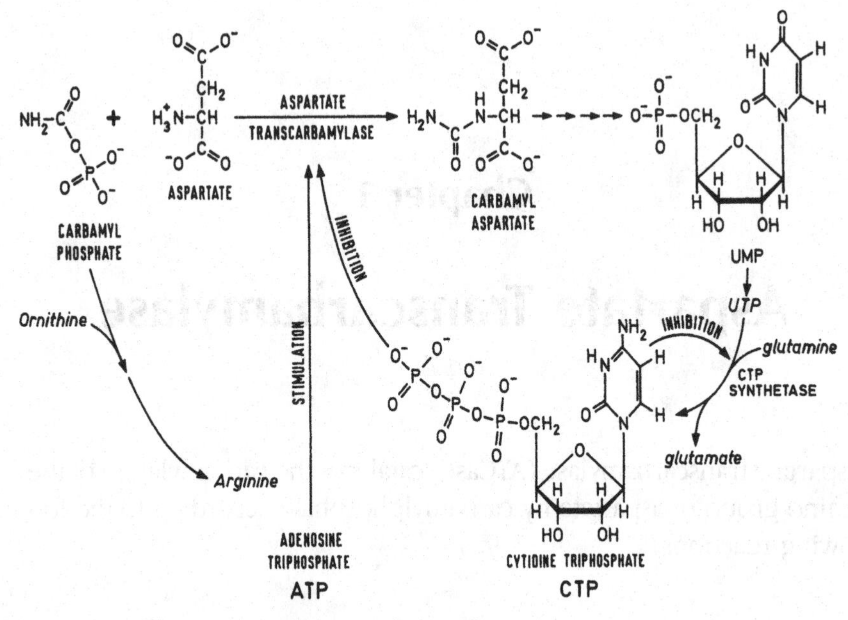

Figure 1. Metabolic regulation of ATCase activity in *E. coli*.

are distinct from the catalytic site. For this reason, they are called allosteric effectors.

The term "allosteric enzyme" applies to enzymes that exhibit feedback inhibition or activation and was extended to enzymes or proteins that show cooperative effects for substrate binding to the catalytic sites. This is the case with ATCase, to which the substrate aspartate binds cooperatively at the catalytic sites. The binding of the first molecule of this substrate facilitates the binding of subsequent molecules. The effectors CTP and ATP bind competitively to the regulatory sites.

1. STRUCTURE OF ATCase

ATCase is a globular protein of molecular mass 305,646 Da and diameter about 100 Å. It is made up of two *catalytic subunits* (relative molecular mass M_r = 101,913) which are trimers of identical *catalytic chains* (M_r = 33,971) and of three *regulatory subunits* (M_r = 33,940) which are dimers of identical *regulatory chains* (M_r = 16,970). Thus, the native enzyme contains six chains of each type (Fig. 2).

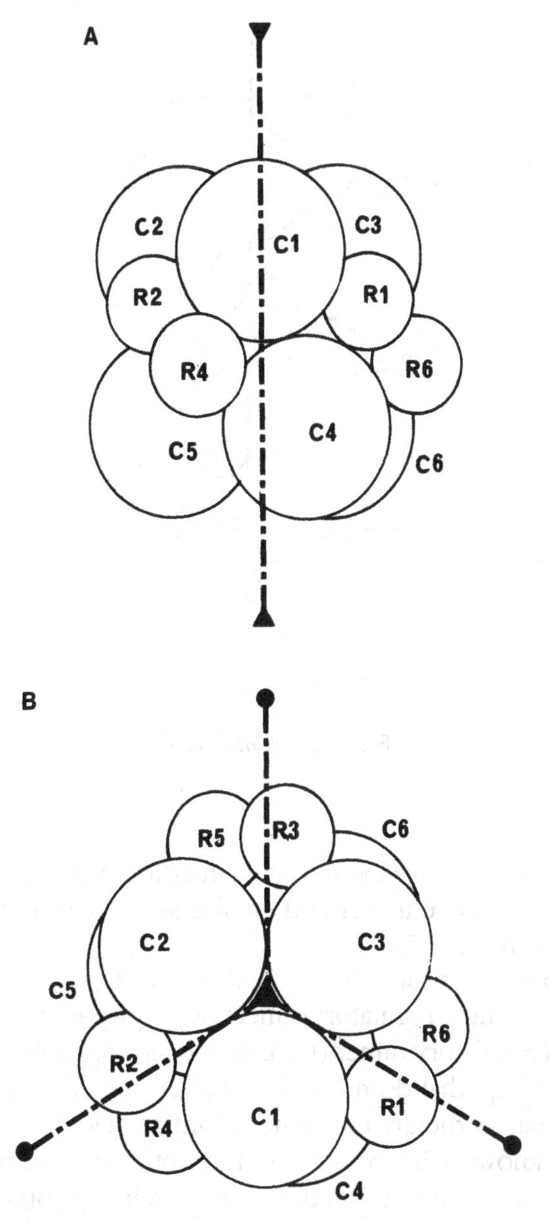

Figure 2. Schematic representation of the quaternary structure of ATCase. The catalytic, C, and regulatory, R, chains are numbered from 1 to 6. (A) View of ATCase perpendicular to the threefold axis. (B) View of ATCase along the threefold axis; the twofold axes of symmetry are indicated. (C) Expanded view of A allowing visualization of the organization of the molecule in trimers of catalytic chains (Cat) and dimers of regulatory chains (Reg). (Adapted from Honzatko *et al.*, 1982, with permission from W. Lipscomb.)

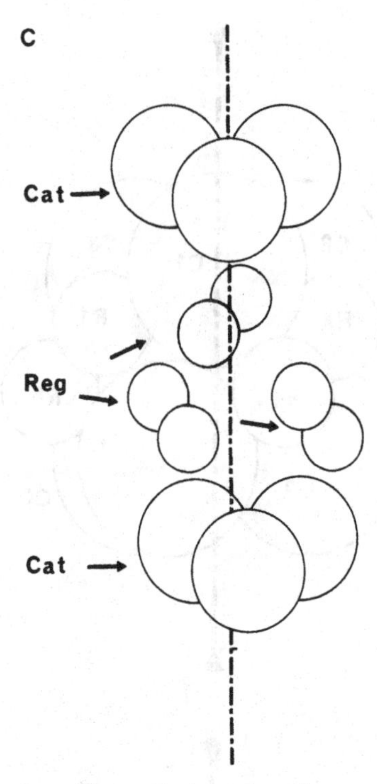

Figure 2. (*continued*)

The catalytic sites are located at the interface between two catalytic chains within the same trimer and involve side chains of amino acids from both chains (Fig. 3).

The two catalytic subunits (C_1-C_2-C_3 and C_4-C_5-C_6) are maintained in contact by the three regulatory dimers (R_1–R_6, R_2–R_4, R_3–R_5). Figure 4 shows how a regulatory dimer (for example, R_1–R_6) links two catalytic chains (C_1 and C_6) that do not belong to the same trimer.

The primary structure (sequence of amino acids) of the two types of chains is known (Fig. 5). Among the noticeable features are the following: the regulatory chain does not contain tryptophan, but contains four cysteine residues, clustered in the C-terminal region of this chain, which play an important structural role, as discussed later. The catalytic chain contains two tryptophan residues whose presence can be used for fluorescence studies (Royer *et al.*, 1987).

The three-dimensional structure of ATCase was determined by X-ray crystallography in W. Lipscomb's laboratory (Harvard University)

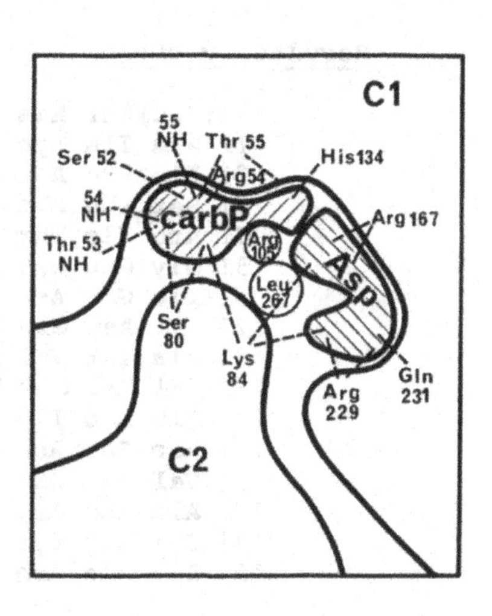

Figure 3. Schematic representation of the active site indicating the amino acids that are involved in substrate binding. CarbP and Asp refer to the approximate positions of the two substrates. The dashed lines indicate the ionic and hydrogen bonds and other polar contacts between groups that are separated by less than 3.5 Å. Numbers associated with the designations for amino acids indicate position in the primary structure. The same is true for some α-amino groups (NH). (Adapted from Volz *et al.*, 1986, with permission from W. Lipscomb.)

and is presently known to a resolution of 2.5 Å. The structure of a catalytic chain-regulatory chain pair is schematically represented in Fig. 6, which also shows the position of the zinc atom that is associated to the regulatory chain R_1. Each regulatory chain contains such a zinc atom, which is linked to the sulfur atoms of the four cysteine residues. The presence of the zinc ensures the correct structural organization of this domain of the regulatory chains, allowing their interaction with the

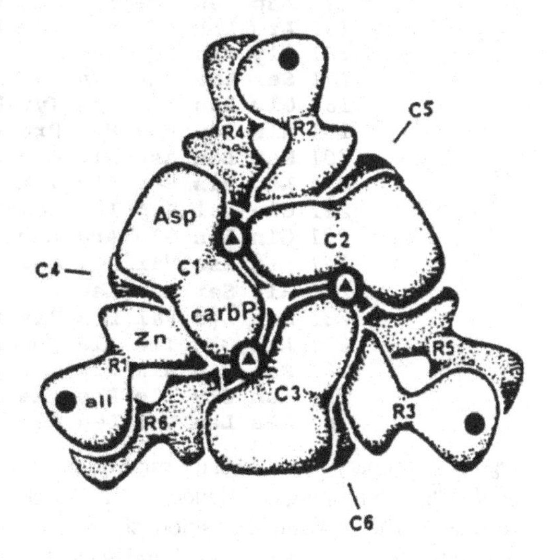

Figure 4. Representation of the quaternary structure of ATCase. ▲, Region of the catalytic site; ●, region of the regulatory site. Asp, Aspartate binding domain of the catalytic chain; carbP, carbamylphosphate binding domain of the catalytic chain; Zn, zinc domain of the regulatory chain; all, allosteric domain of the regulatory chain where the effector binding sites are localized. (Adapted from Krause *et al.*, 1987, with permission from W. Lipscomb.)

Regulatory chain

1 (Met) Thr His Asp Asn Lys Leu Gln Val Glu
11 Ala Ile Lys Arg Gly Thr Val Ile Asp His
21 Ile Pro Ala Gln Ile Gly Phe Lys Leu Leu
31 Ser Leu Phe Lys Leu Thr Glu Thr Asp Gln
41 Arg Ile Thr Ile Gly Leu Asn Leu Pro Ser
51 Gly Glu Met Gly Arg Lys Asp Leu Ile Lys
61 Ile Glu Asn Thr Phe Leu Ser Glu Asp Gln
71 Val Asp Gln Leu Ala Leu Tyr Ala Pro Gln
81 Ala Thr Val Asn Arg Ile Asp Asn Tyr Glu
91 Val Val Gly Lys Ser Arg Pro Ser Leu Pro
101 Glu Arg Ile Asp Asn Val Leu Val Cys Pro
111 Asn Ser Asn Cys Ile Ser His Ala Glu Pro
121 Val Ser Ser Ser Phe Ala Val Arg Lys Arg
131 Ala Asn Asp Ile Ala Leu Lys Cys Lys Tyr
141 Cys Glu Lys Glu Phe Ser His Asn Val Val
151 Leu Ala Asn

Catalytic chain

1 Ala Asn Pro Leu Tyr Gln Lys His Ile Ile
11 Ser Ile Asn Asp Leu Ser Arg Asp Asp Leu
21 Asn Leu Val Leu Ala Thr Ala Ala Lys Leu
31 Lys Ala Asn Pro Gln Pro Glu Leu Leu Lys
41 His Lys Val Ile Ala Ser Cys Phe Phe Glu
51 Ala Ser Thr Arg Thr Arg Leu Ser Phe Glu
61 Thr Ser Met His Arg Leu Gly Ala Ser Val
71 Val Gly Phe Ser Asp Ser Ala Asn Thr Ser
81 Leu Gly Lys Lys Gly Glu Thr Leu Ala Asn
91 Thr Ile Ser Val Ile Ser Thr Tyr Val Asp
101 Ala Ile Val Met Arg His Pro Gln Glu Gly
111 Ala Ala Arg Leu Ala Thr Glu Phe Ser Gly
121 Asn Val Pro Val Leu Asn Ala Gly Asp Gly
131 Ser Asn Gln His Pro Thr Gln Thr Leu Leu
141 Asp Leu Phe Thr Ile Gln Glu Thr Gln Gly
151 Arg Leu Asp Asn Leu His Val Ala Met Val
161 Gly Asp Leu Lys Tyr Gly Arg Thr Val His
171 Ser Leu Thr Gln Ala Leu Ala Lys Phe Asp
181 Gly Asn Arg Phe Tyr Phe Ile Ala Pro Asp
191 Ala Leu Ala Met Pro Gln Tyr Ile Leu Asp
201 Met Leu Asp Glu Lys Gly Ile Ala Trp Ser
211 Leu His Ser Ser Ile Glu Glu Val Met Ala
221 Glu Val Asp Ile Leu Tyr Met Thr Arg Val
231 Gln Lys Glu Arg Leu Asp Pro Ser Glu Tyr
241 Ala Asn Val Lys Ala Gln Phe Val Leu Arg
251 Ala Ser Asp Leu His Asn Ala Lys Ala Asn
261 Met Lys Val Leu His Pro Leu Pro Arg Val
271 Asp Glu Ile Ala Thr Asp Val Asp Lys Thr
281 Pro His Ala Trp Tyr Phe Gln Gln Ala Gly
291 Asn Gly Ile Phe Ala Arg Gln Ala Leu Leu
301 Ala Leu Val Leu Asn Arg Asp Leu Val Leu

Figure 5. Primary structure of the catalytic and regulatory chains (Weber, 1968; Hoover *et al.*, 1983; Konigsberg and Henderson, 1983; Schachman *et al.*, 1984). The methionine residue in the *N*-terminal position of the regulatory chain is removed during the maturation of the enzyme (Hervé and Stark, 1967).

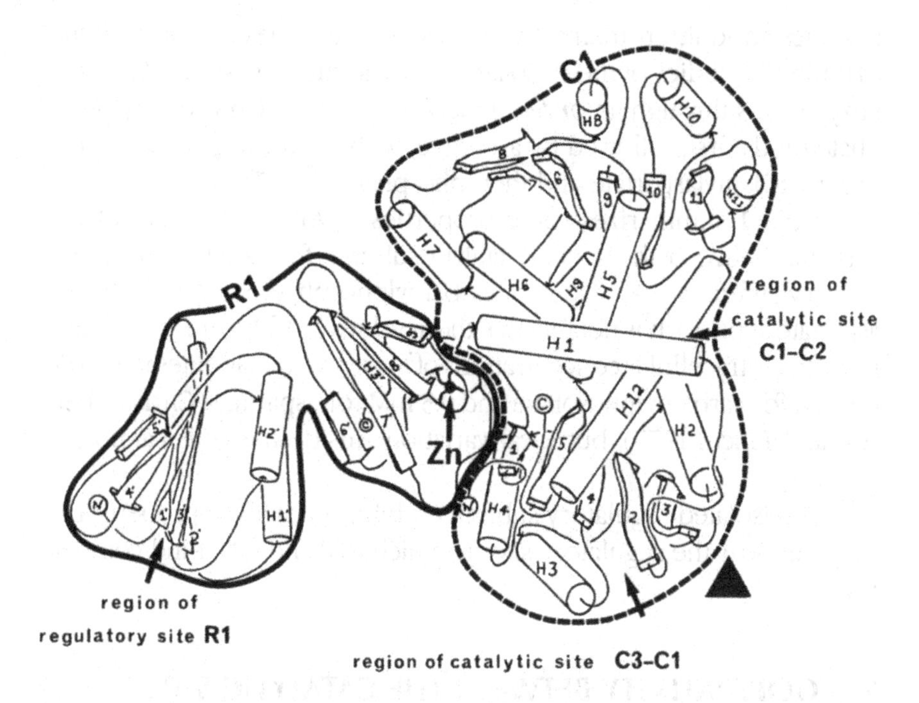

Figure 6. Schematic representation of the structure of a catalytic chain–regulatory chain pair. Cylinders refer to α-helical structures; arrows indicate regions that are organized in β sheets. (Adapted from Ke *et al.*, 1988, with permission from W. Lipscomb.)

catalytic chains. This is necessary for the subunit association. In fact, as discussed later, it is the reaction of some mercurial reagents with the cysteine residues which allows the experimental dissociation of the enzyme into catalytic and regulatory subunits. Therefore, ATCase is not strictly speaking a metalloenzyme, since zinc plays no role in the catalytic property of the enzyme but rather plays a purely structural role with respect to its regulatory properties.

Under the influence of mercurial reagents such as *p*-hydroxymercuribenzoate, ATCase can be dissociated into catalytic and regulatory subunits, which can then be separated. This dissociation is reversible, and under appropriate conditions, ATCase can be reconstituted from its subunits. The isolated catalytic subunits are still able to catalyze the carbamylation of the amino group of aspartate, but they do not exhibit the regulatory properties of the native enzyme. They show no cooperative interactions between the catalytic sites, and their saturation curve by aspartate is hyperbolic, a property characteristic of Michaelian enzymes. In addition, their activity is insensitive to the effectors CTP and ATP. Kinetic studies of these isolated catalytic subunits showed that the enzyme acts through an *ordered mechanism* in which carbamylphosphate binds first, followed by aspartate, with products released in the order carbamylaspartate and then phosphate (Fig. 7).

Table 1 summarizes some properties of ATCase and its isolated subunits. The K_M of the isolated catalytic subunits for their two substrates are very different (5×10^{-5} M for carbamylphosphate and 2×10^{-2} M for aspartate at pH 8). It is notable that these values are far from the average reported intracellular concentrations of these two metabolites in *E. coli*: 0.8 mM for carbamylphosphate and 0.3 mM for aspartate (Christopherson and Finch, 1977), but local variations inside the cells cannot be excluded.

The isolated regulatory subunits exhibit no catalytic property, but they still bear the regulatory sites at which ATP and CTP bind competitively.

2. COOPERATIVITY BETWEEN THE CATALYTIC SITES

ATCase exhibits cooperative interactions between the catalytic sites for the utilization of aspartate. The binding of the first molecules of this substrate to the enzyme–carbamylphosphate complex facilitates the binding of the following ones. Such interactions are called *homotropic* (between identical sites). As a consequence, the graphical representation of the rate of reaction versus aspartate concentration is sigmoidal. The binding of carbamylphosphate, the first substrate of ATCase, takes place without significant cooperativity.

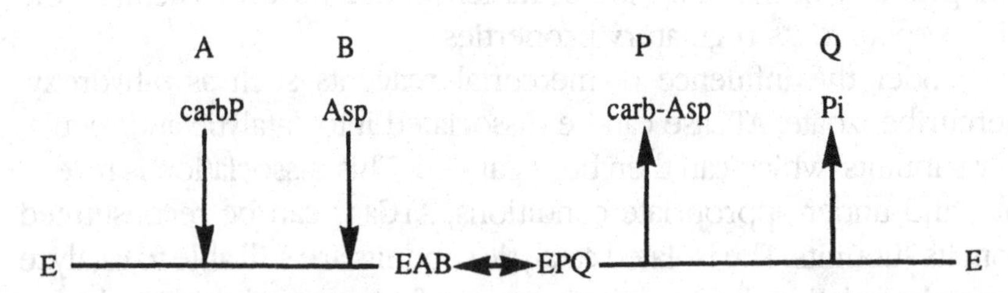

Figure 7. Enzymatic mechanism of *E. coli* ATCase. The K_Ms of the isolated catalytic subunits for their substrates at pH 7 and 30°C are as follows: carbamylphosphate, 0.03 mM; aspartate, 20 mM; carbamylaspartate, 20 mM; phosphate, 4.2 mM (Porter *et al.*, 1969; Hsuanyu and Wedler, 1987).

Table 1. Properties of ATCase and Its Isolated Subunits[a,b]

Properties	ATCase (E)	Cat.	Reg.
M_r[c]	305,646	101,913	33,940
A_{280} (1 mg/ml), pH 7, l = 1 cm	0.59	0.72	0.35
A_{280}/A_{260}	1.82	1.98	1.2
Number of cysteine residues	30	3	8
CTP required for ½ saturation (M)[d]	1.2×10^{-5}		5×10^{-5}
Percent inhibition by 5×10^{-4} CTP in the presence of 5×10^{-3} M ASP	75	<3	
$S_{0.5}$ ASP (M)[e]	13×10^{-3}	20×10^{-3}	
Specific activity[f]	8×10^3	13×10^3	<30

[a]Adapted from Gerhart and Holoubek, 1967, and updated.
[b]Abbreviations: E, the native ATCase [$2(C_3)3(R_2)$]; Cat., the isolated catalytic subunits [trimer (C_3)]; Reg., the isolated regulatory subunits [dimer (R_2)].
[c]Values calculated without the *N*-terminal methionines (Fig. 5).
[d]Under the following standard conditions: 28°C, 40-mM phosphate buffer (pH 7), 15-mM aspartate, and 4-mM carbamylphosphate.
[e]At 37°C in the presence of 50-mM Tris–acetate buffer (pH 8) and 5-mM carbamylphosphate.
[f]Expressed as micromoles of carbamylaspartate synthesized per hour per milligram of enzyme under the conditions indicated under footnote *d*.

These homotropic interactions between catalytic sites can be explained by a shift of the enzyme from a conformation having a weak affinity for aspartate (*T* state) to a conformation with a high affinity for this substrate (*R* state). These two extreme conformations of ATCase are known to a resolution of 2.5 Å. Figure 8 shows the α-carbon backbone of these two extreme conformations as they were determined by X-ray crystallography in the laboratory of W. Lipscomb (Honzatko *et al.*, 1982; Krause *et al.*, 1987).

During the transition of *T* to *R*, the two catalytic subunits move apart by 12 Å along the threefold axis of symmetry at the same time as they rotate by 5° each around this axis. This movement is accompanied by a rearrangement of the regulatory subunits toward the center of the molecule. This quaternary structure change is coupled to a change in the tertiary structure of the catalytic subunit. In each catalytic chain, the carbamylphosphate domain and the aspartate domain come closer together, leading to the formation of a more compact and efficient catalytic site.

This structural transition involves modifications of the ionic interactions between the subunits. Figure 9 shows the nature of some of these interactions in the two extreme conformations of the enzyme (Ladjimi and Kantrowitz, 1988).

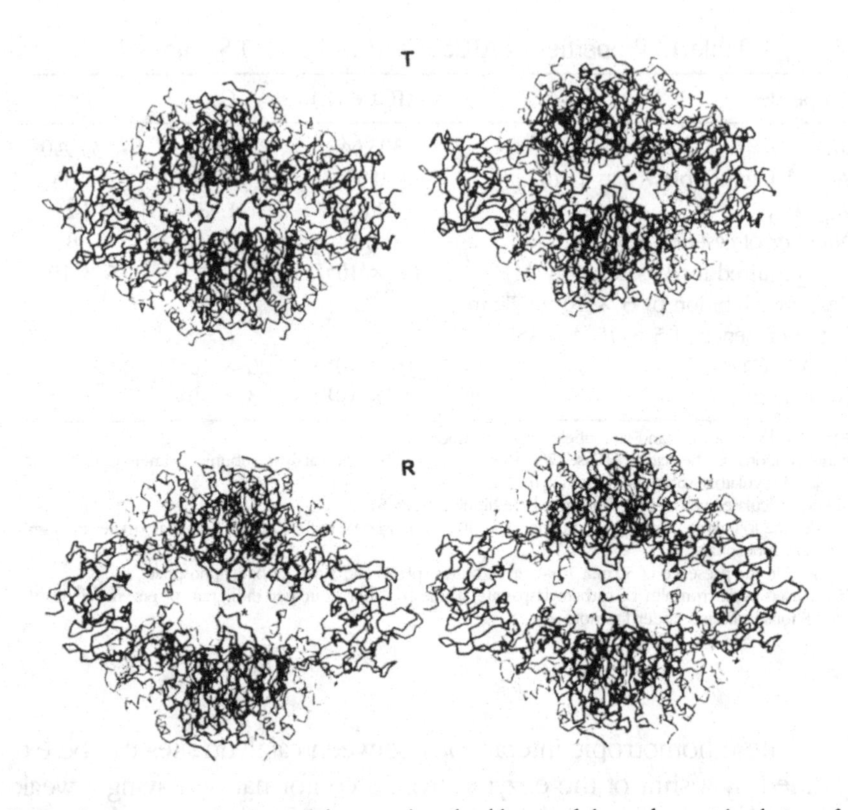

Figure 8. Stereoscopic view of the α-carbon backbone of the polypeptide chains of ATCase in the *T* and *R* conformations. (From Krause *et al.*, 1987, with permission from W. Lipscomb.)

The pH dependence of the reaction catalyzed by ATCase changes when the concentration of the substrate aspartate is raised, suggesting that the two *T* and *R* states differ by the pK_a(s) of one or several groups involved in aspartate binding or catalysis. The isolated catalytic subunits do not exhibit such a variation, and analysis of their pH dependence for activity allows the characterization of three pK_as involved in substrate binding and catalysis (Fig. 10). The residue histidine 134 (Fig. 3) is directly implicated in the binding of carbamylphosphate. However, the deprotonation of another unknown residue is involved in catalysis, a result consistent with the proposed mechanism for ATCase activity. In this mechanism, it is postulated that the interaction between a group that belongs to the enzyme and the carbonyl group of carbamylphosphate increases the polarization of this carbonyl group, facilitating its nucleophilic attack by the amino group of aspartate.

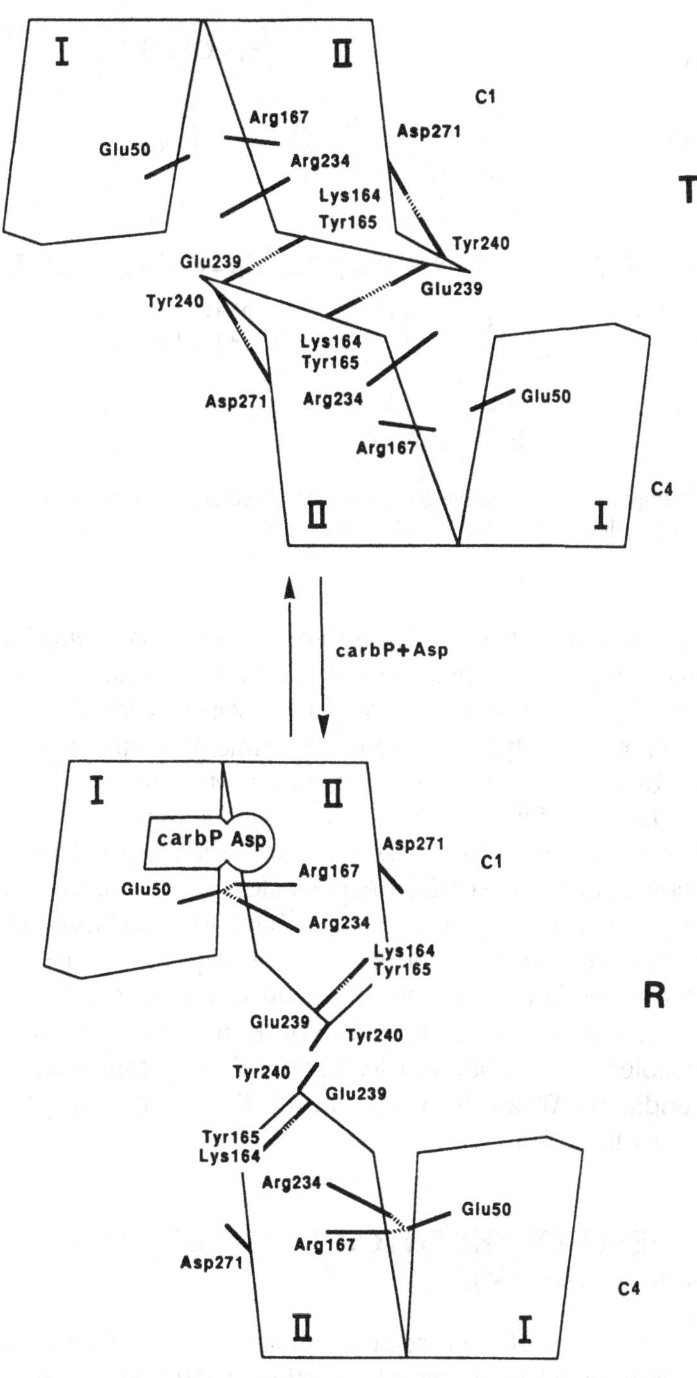

Figure 9. Changes in ionic interactions between catalytic chains during the allosteric transition. I, Carbamylphosphate binding domain; II, aspartate binding domain. (Adapted from Ladjimi and Kantrowitz, 1988, with permission from M. Ladjimi.)

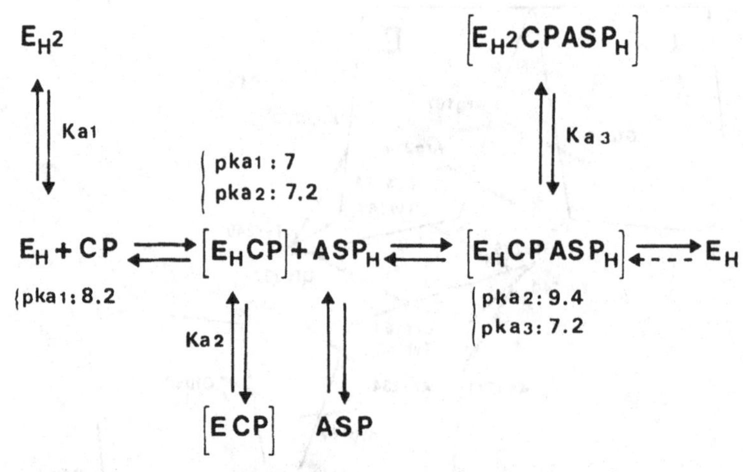

Figure 10. Implication of ionizable groups of the isolated catalytic subunits in substrate binding and catalysis (Léger and Hervé, 1988).

The *T*-to-*R* transition in ATCase appears to be *concerted* in the sense that it is complete before the saturation of its catalytic sites by aspartate or its analogs. In other words, the *R* state is reached before the filling of all of the catalytic sites. Some experimental results suggest that substrate binding to only one of the six catalytic sites is sufficient to promote the shift of the entire enzyme molecule to the *R* state (Foote and Schachman, 1985). This is taken into account in Fig. 9. However, if we consider a population of ATCase molecules in solution, aspartate (or its analogs) will tend to preferentially bind to those molecules of enzyme that are already in the *R* state. The experiments performed under these conditions indicate that, statistically, four molecules of substrate or analog must be bound in order to ensure that all of the enzyme molecules have bound at least one molecule of substrate. Under these conditions ATCase is entirely in the *R* state, although it is not saturated by the substrate.

3. INFLUENCE OF THE EFFECTORS CTP, UTP, AND ATP (PSE MECHANISM)

The activity of ATCase is regulated by the effectors CTP and ATP, a phenomenon called *heterotropic* interactions. CTP (feedback inhibitor) and ATP (activator) bind reversibly and competitively to the regulatory sites (Figs. 4 and 6). UTP has no effect on the enzyme activity alone, but is acts synergistically with CTP (Wild *et al.*, 1989). The first models that were proposed to account for the influence of effectors on allosteric enzymes postulated that they act by shifting the $T \rightleftharpoons R$ equilibrium involved in the homotropic cooperative interaction between the catalytic sites, as the substrate does. If such a mechanism applies to ATCase, ATP would displace the equilibrium toward the *R* state and CTP would displace it toward the *T* state. In fact, it appears that in this enzyme, the effectors act through a mechanism that is distinct from that involved in the homotropic interactions between the catalytic sites. Figure 11 depicts

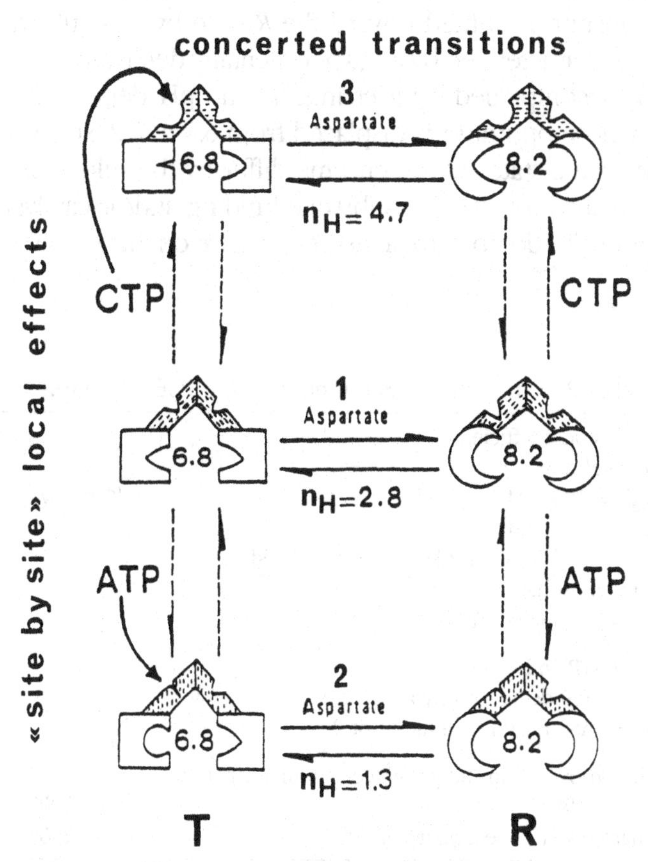

Figure 11. Mechanism of heterotropic interactions in ATCase (PSE mechanism). n_H is the Hill coefficient (see Chapter 5, Section 1) for each of the transitions; 6.8 and 8.2 are the optimum pHs for activity of the different conformations of the enzyme (Thiry and Hervé, 1978; Tauc *et al.*, 1982).

the alternative mechanism that was proposed (primary–secondary effects, or PSE mechanism) to describe the heterotropic interactions. Originally based on purely biochemical evidence (Thiry and Hervé, 1978; Tauc *et al.*, 1982), this mechanism was later confirmed by X-ray solution scattering experiments (Hervé *et al.*, 1985) and equilibrium isotope exchange kinetics (Hsuanyu and Wedler, 1988).

In this scheme (Fig. 11), ATCase is represented in a very schematic way as two catalytic chains linked by a dimer of regulatory chains. On the left of the figure, the enzyme is in the *T* state of low affinity for aspartate; on the right side, it is in the *R* state of high affinity for this substrate. Equilibrium 1 accounts for the homotropic cooperative effects between the catalytic sites for the binding of aspartate in the absence of effectors. This equilibrium is shifted toward the *R* state by this substrate. This transition is characterized by an experimentally determined Hill coefficient. It is accompanied by a change in the pH dependence of the reaction, whose optimum is displaced from 6.8 to 8.2, suggesting that the two extreme states of the enzyme differ by the values of pK_a of a group or groups involved in substrate binding and/or catalysis.

ATP and CTP do not have a *direct* influence on this equilibrium but

Table 2. Quantitative Parameters of the PSE Mechanism

Non-exclusion coefficient[a]	
Aspartate[b]	
$K_M^T \cong 130$ mM	$1/C = K_T/K_R \cong 20$
$K_M^R \cong$ 6 mM	
$\Delta G = -1.8$ kcal/mol $= -7.5$ kJ/mol	
Carbamylphosphate	
$1/C = 2$ to 3	
Effects of ATP and CTP	
ATP: K_M ASP decreased by a factor of 2	
CTP: K_M ASP raised by a factor of 2	
Stimulation by ATP in the presence of 1-mM aspartate	
Primary effect	100%
Primary effect + secondary effect	250%

[a]The nonexclusion coefficient *C* is the ratio of the affinities of the *T* and *R* states for the substrate (Chapter 5). In the case of aspartate, the ratio of K_M values is approximated to the ratio of dissociation constants K_T and K_R. $\Delta G = -2.3RT \log K_T/K_R$.

[b]In the presence of 20-mM Tris–acetate buffer (pH 8.5)–4-mM carbamylphosphate. The percentage of stimulation by ATP is calculated as indicated Chapter 5, Section 2.1.

act according to what was called the primary effect and the secondary effect. Through a local conformational change that can be detected by spectroscopic analysis, ATP binding induces an increase in the affinity for aspartate of the corresponding catalytic site (*primary effect*). In the presence of a given concentration of aspartate, this, in turn, induces an increase in the degree of occupancy of the catalytic sites by this substrate, the consequence of which is a shift of the equilibrium $T \rightleftharpoons R$ toward the R state (*secondary effect*, equilibrium 2). CTP has the opposite effect (secondary effect, equilibrium 3), lowering the affinity of the catalytic sites for aspartate (*primary effect*). However, in the case of CTP a slight additional direct effect on the $T \rightleftharpoons R$ equilibrium has been detected (Hervé *et al.*, 1985). The effect of these nucleotides on the rate of reaction is exerted "site by site"; i.e., this effect varies linearly with the number of regulatory sites to which the effectors are bound. The available quantitative parameters of this mechanism are presented in Table 2.

Investigation of these phenomena by equilibrium isotope exchange kinetics has shown that the primary effects of the nucleotides specifically alter the rate constant for the binding of aspartate to the catalytic sites (Hsuanyu and Wedler, 1988).

More detailed reviews on ATCase were recently published (Allewell, 1989; Hervé, 1989).

Chapter 2

Molecular Genetics

Regulation of Aspartate Transcarbamylase Biosynthesis

1. THE ATCase OPERON AND ITS REGULATION

The catalytic and regulatory chains are coded for by two genes (*pyrB* and *pyrI*, respectively) that constitute an operon located at 96′ on the *E. coli* chromosome. Their biosynthesis is therefore directed by a polycistronic mRNA that is expressed in the order catalytic chain and then regulatory chain (Perbal *et al.*, 1977; Roof *et al.*, 1982). Such an organization ensures a balanced production of the two types of chains, whose association constitutes ATCase. The 3′-terminal end of this mRNA codes for a third protein, called X, whose role is unknown (J. Wild, personal communication). The expression of this operon is regulated by the intracellular level of UTP, a pyrimidine nucleotide whose biosynthesis is initiated by ATCase (Fig. 1). When UTP is present in the cell at a concentration lower than 0.2 mM, the expression of this operon increases greatly.

Expression of the ATCase operon is regulated through an *attenuation mechanism*, the principle of which is illustrated in Fig. 12 (Roland *et al.*, 1985). Upstream from the structural genes coding for the catalytic and regulatory chains one finds the *promoter* and, on the coding strand, a region that is rich in purine bases called the *transcription pause site*. Between this site and the first structural gene there is a region called the *attenuator*, containing a transcription termination *hairpin* that is made of two complementary sequences belonging to the same DNA strand. The mRNA transcribed from this region can fold into a loop in which

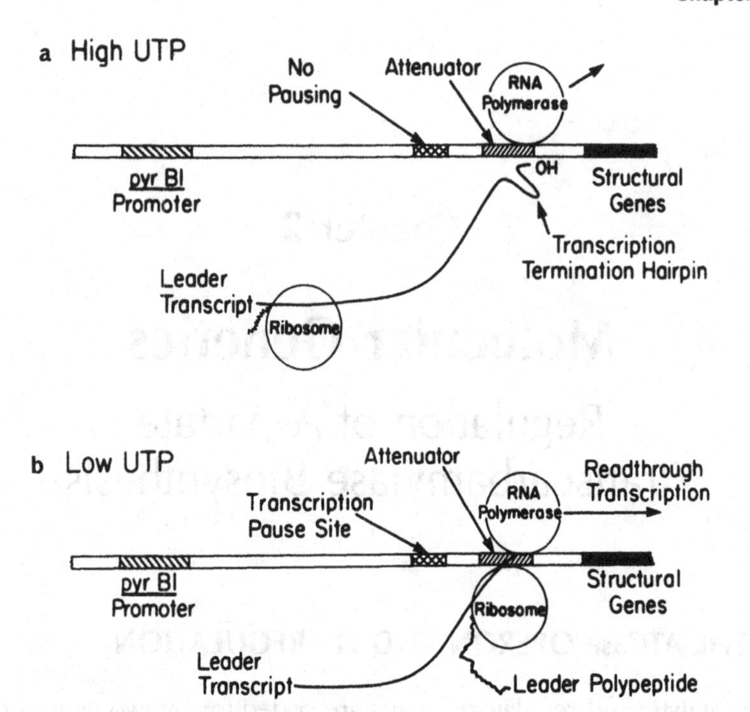

Figure 12. Model for the regulation of *pyrBI* operon expression. (Adapted from Roland *et al.*, 1985, with permission of C. L. Turnbough.)

these regions are associated by base pairing. This sequential organization allows the regulation of operon expression in the following way (Fig. 12).

(a) At high intracellular UTP concentrations, RNA polymerase binds to the promoter and transcribes the leader region at a rate which is not affected by the transcription pause site; under these conditions, there is enough time for formation of the hairpin before the arrival of the first ribosomes, which translate this leader sequence. The hairpin is a transcription termination signal that induces the release of the RNA polymerase before transcription of the structural genes (*repression*).

(b) At low intracellular UTP concentrations, RNA polymerase binds to the promoter and transcribes the leader region. At the level of the purine-rich transcription pause site, the polymerase is slowed down because of the lack of pyrimidine nucleotides. Under these conditions, the ribosomes can closely follow the RNA polymerase (transcription–translation coupling), preventing the formation of the hairpin structure, and RNA polymerase reads through the structural genes (*derepression*).

2. EXPERIMENTAL DETERMINATIONS: MEASUREMENT OF THE LAGTIME FOR DEREPRESSION AND REPRESSION OF ATCase OPERON EXPRESSION

2.1. Principle

E. coli cells grown in the presence of uracil are rapidly resuspended in the same growth medium but deprived of this metabolite (t_0 of derepression). At various times, samples of cells are taken and their ATCase activity is measured. A second cell preparation is resuspended in the uracil-deprived medium, but after some time this metabolite is added (t_0 of repression). The ATCase activity is determined in these cells before and after this addition.

2.2. Bacterial Strain

The bacterial strain used in these experiments has a mutation that ensures a high level of expression of the ATCase operon. This type of strain is obtained as follows (Gerhart and Holoubek, 1967). The first step consists of the selection after mutagenesis of bacteria that are auxotrophic for pyrimidine as a consequence of a point mutation in the *pyrF* gene, which codes for *orotate decarboxylase*, an enzyme that in *E. coli* catalyzes the fourth reaction of the pyrimidine pathway. In the next step, partial revertants, whose growth is limited by the activity of this enzyme (*pyrF* bradytrophs), are selected on petri dishes containing the uracil-deprived medium. Such mutants synthesize enough pyrimidine nucleotides to allow the synthesis of the RNA coding for ATCase but not enough to provoke the repression of ATCase operon expression (Fig. 12). Some strains are now available in which the ATCase operon is carried by a multicopy plasmid. The presence of such plasmids in a *pyrF* bradytroph strain allows the biosynthesis of large amounts of enzyme.

2.3. Growth Medium

The growth medium used contains (in grams per liter):

1. disodium phosphate: 8.7
2. monopotassium phosphate: 5.3
3. ammonium sulfate·$4H_2O$: 0.10

4. calcium nitrate: 0.005
5. zinc sulfate·$7H_2O$: 0.005
6. iron(III) sulfate·$7H_2O$: 0.005

The pH is adjusted to 7; after sterilization, a sterile solution of glucose is added to a final concentration of 4g/liter.

Bacteria repressed for the biosynthesis of ATCase are obtained by growth in the same medium but containing 50 μg/ml uracil.

2.4. Experimental Procedures

Equipment

- Shaking water bath at 37°C
- Sonicator for the preparation of cell extracts
- Millipore apparatus for filtration under vacuum and 45-mm membranes (pore size, 1.2 μm)
- 250-ml culture flasks containing sterile growth medium. These flasks may have a side arm allowing the direct determination of optical density (Fig. 13)
- Spectrophotometer for visible light with a cuvette holder able to receive the flask side arm

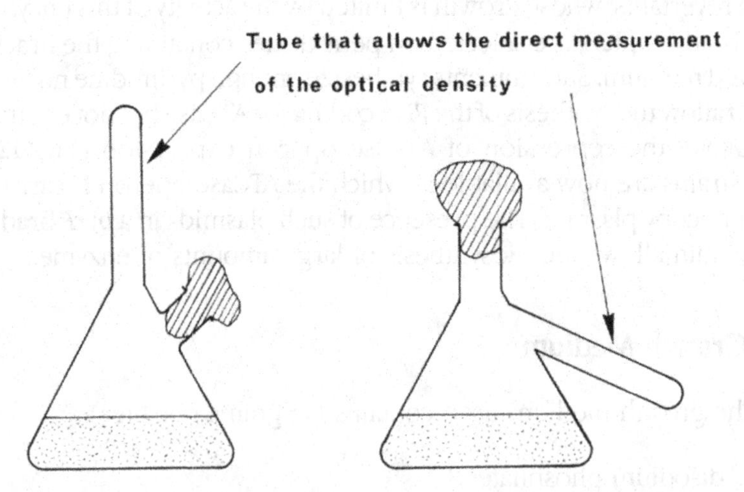

Figure 13. Types of flask allowing direct measurement of the optical density of bacterial cultures. This measurement is made by tilting the flask to fill the tube.

- Tabletop centrifuge for test tubes
- Oven at 37°C

Derepression Experiments

Starting from a preculture of *E. coli*, seed a 50-ml sample of the growth medium containing 30 μg/ml uracil. Incubate the culture at 37°C in a shaking water bath. When the optical density (turbidity) of the culture is about 0.4 at 546 nm, filter rapidly, using the Millipore filtration system previously warmed to 37°C in an oven. Rapidly wash the bacteria directly on the filter with 50 ml of the growth medium (without uracil) prewarmed to 37°C. Resuspend the bacteria as quickly as possible in 30 ml of the same medium; the filter is thrown into the medium. The entire operation must not take more than 1 min. Remove 900-μl samples of bacterial suspension every 30 sec and rapidly transfer them into a small test tube containing 100 μl of a solution of chloramphenicol at 100 μg/ml (to inhibit protein biosynthesis) and cooled at 0°C. After 6 min of de-repression, the samples should be taken every minute. At the end of this incubation, centrifuge the samples and resuspend the cell pellet in 1 ml of 40-mM Tris–acetate buffer (pH 8)–1-mM β-mercaptoethanol–0.1-mM EDTA. Treat these samples three times for 30 sec each time with a sonicator to obtain the cell extracts. Determine the ATCase activity as described in Chapter 5, using 100–200 μl of each extract. Measure the protein concentration in the first and the last samples by the colorimetric assay described in Chapter 4, using 10–50 μl of extract. Do not forget blanks and the standard curve.

Results

Represent graphically the variation of ATCase activity as a function of time (Fig. 14).

Express the initial and final ATCase activities in term of specific activity (micro-moles of carbamylaspartate formed per hour per milli-gram of protein under the conditions used).

Question 1. *On the basis of an ATCase specific activity of 3000 μmole of carbamylaspartate formed per hour per milligram of ATCase under the conditions described in Chapter 5, calculate the number of enzyme molecules synthesized per minute per cell, knowing that an optical density of 0.5 measured at 546 nm corresponds to about 7 × 10^8 bacteria per ml.*

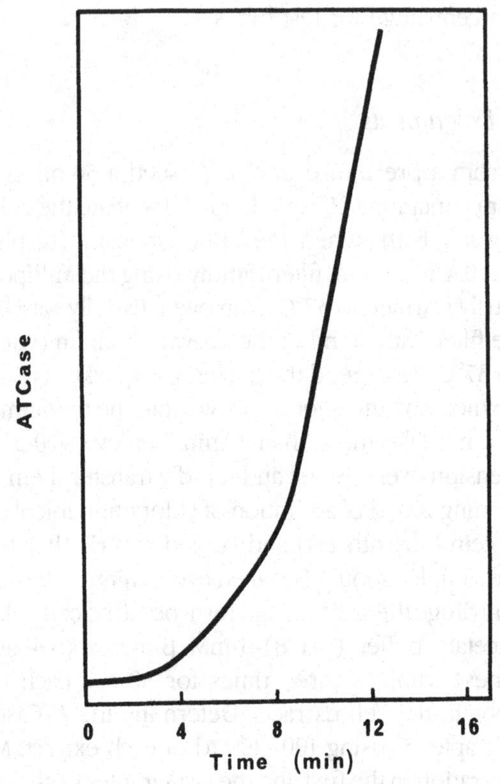

Figure 14. Derepression of the ATCase operon.

Repression Experiments

A bacterial culture is derepressed during 20 min as described above. After 20 min, add rapidly to this culture 5 ml of a uracil solution at 0.5 mg/ml prewarmed to 37°C. Remove a 1-ml sample each minute, from 10 min before uracil addition to 12 min after. Treat these samples as previously described and determine their ATCase activity, using 50-μl aliquots.

Results

Plot the variation of ATCase activity as a function of time (Fig. 15).

Determination of the Half-Life of the Messenger-RNA Coding for ATCase

Grow a 30-ml culture of the bacterial strain in growth medium containing 30 μg/ml uracil (repression) to an optical density of 0.4.

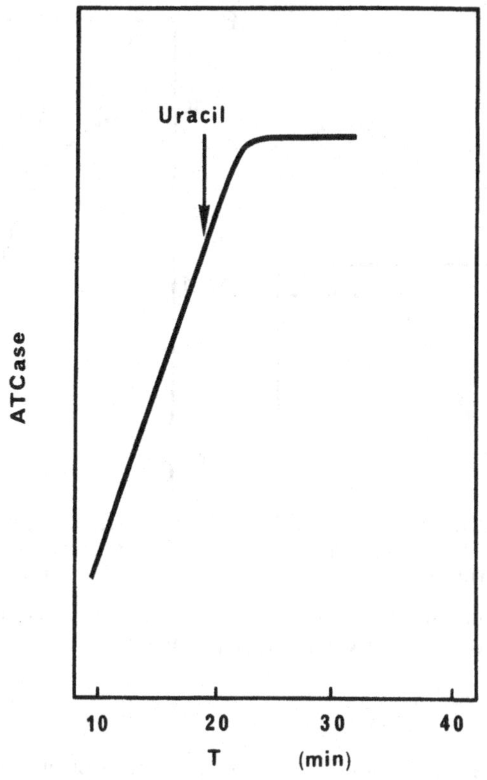

Figure 15. Repression of the ATCase operon.

Rapidly filter the bacteria as previously described and wash them directly onto the filter with 15 ml of uracil-deficient growth medium prewarmed to 37°C. Resuspend the cells in 30 ml of the same medium. These operations should not take more than 1 min. After 1 min of incubation under these conditions, rapidly add 1 ml of a solution of uracil at 100 μg/ml prewarmed to 37°C. Thus, the bacteria are exposed to the conditions of derepression for 1 min. This is called an elementary wave of derepression. Immediately after adding uracil, remove 1-ml samples every 30 sec for 15 min. Treat these samples as previously described for determination of ATCase activity, using aliquots of 50–100 μl.

Results

Plot the variation of ATCase activity as a function of time. This activity approaches a constant value E_∞. Using semilogarithmic graph paper,

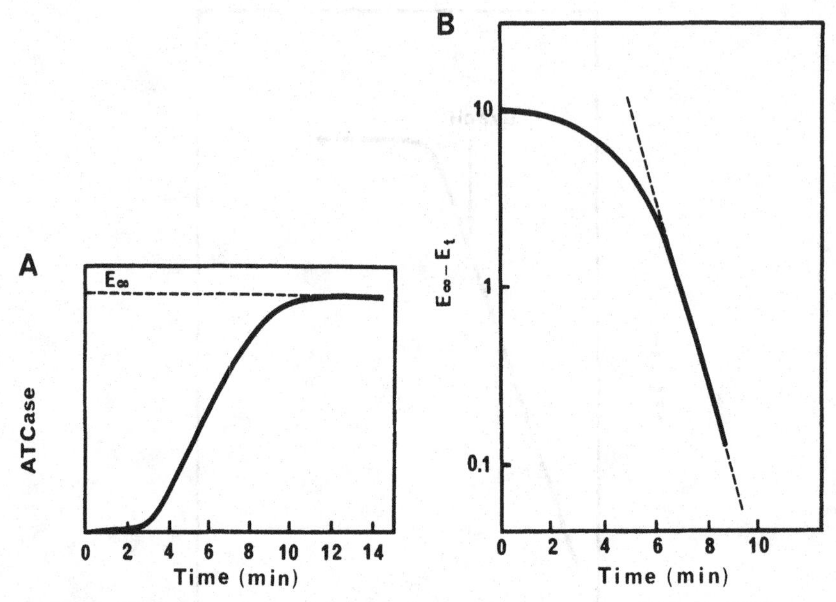

Figure 16. Elementary wave of derepression (A) and estimation of the half-life of the mRNA coding for ATCase (B) (Perbal and Hervé, 1972).

represent the evolution of $E_\infty - E_t$ as a function of time (Fig. 16). Deduce from this graph the half-life of the RNA coding for ATCase. This is done by reading the time that is necessary to decrease $E_\infty - E_t$ by a factor of 2 in the linear part of the curve.

Question 2. *What hypothesis underlies this method of calculation of the half-life of the mRNA?*

Chapter 3

Purification of Aspartate Transcarbamylase and Its Subunits

ATCase is purified from an overproducing strain of *E. coli* of the type described in Chapter 2. One may recall that because of a mutation on the orotate decarboxylase *pyrF* gene, these bacteria synthesize sufficient pyrimidine nucleotides to allow the biosynthesis of the mRNA coding for ATCase, but not enough to repress expression of the corresponding operon. To grow these bacteria with a good yield, one seeds a glucose minimum medium (Chapter 2) containing a limiting amount of uracil (4 μg/ml). Growth of these bacteria and biosynthesis of ATCase take place in three phases (Fig. 17).

During the first phase, growth is fast and there is virtually no synthesis of ATCase. During the second phase, the exhaustion of uracil leads to a slowing down of growth, which is then limited by the residual activity of the orotate decarboxylase and to an important and continuous expression of the ATCase operon. Cells are collected by centrifugation when the culture has reached the stationary phase (Fig. 17). Whatever its volume, such a culture is seeded with a few percent of a preculture obtained in the presence of 30-μg/ml uracil.

1. PURIFICATION OF ATCase

ATCase is purified according to the method of Gerhart and Holoubek (1967). This procedure is described here for the treatment of about 300 g (wet wt) of bacteria. At the end of phase 2 of the culture (Fig. 17),

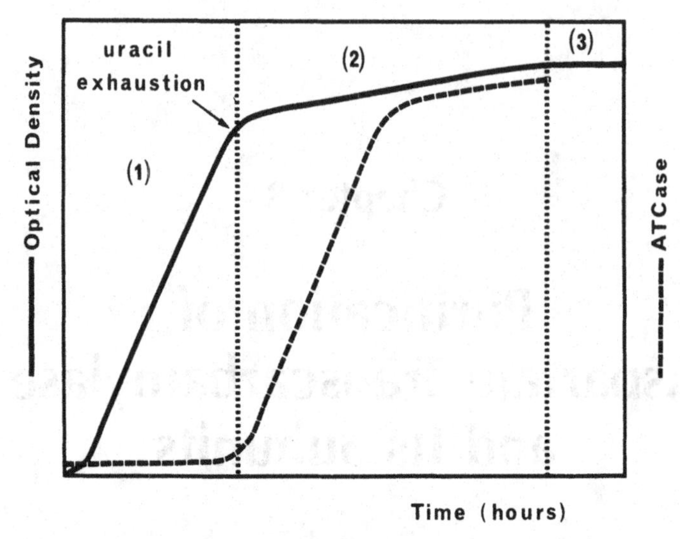

Figure 17. Cell growth and biosynthesis of ATCase in *pyrF* bradytroph strains in the presence of limiting uracil. Cell growth is monitored on the basis of the optical density at 546 nm. (1) Rapid exponential growth phase under conditions of repression (4 μg/ml uracil); (2) slow exponential growth phase after exhaustion of the exogenous uracil and derepression of ATCase operon; (3) stationary phase at the end of growth. ATCase activity is expressed as units per milligram of protein. (From Gerhart and Holoubek, 1967, with permission of J. C. Gerhart.)

the bacteria are collected by centrifugation (15 min at 3000 *g*). Starting with the strains previously described, these conditions yield about 350 mg of enzyme. For treatment of different amounts of bacteria, the volumes of reagents indicated below should be changed proportionally.

1.1. Preparation of the Bacterial Extract

The cells are suspended in 900 ml of 0.1-M Tris–HCl buffer (pH 9.2) and disrupted in either a sonicator or a Manton–Gaulin or French Press, in which the cell walls are broken through the alternation of pressure and depressure. Alternatively, one can use a Waring blender with small glass beads.

1.2. Heat Step

To each liter of extract add 350 ml of a 3.6-M solution of ammonium sulfate prepared in 0.1-M Tris–HCl buffer (pH 9.2). Fractions (400 ml) of the mixture are incubated for 10 min at 75°C in a water bath with stirring and then quickly chilled in ice water. Under these conditions, an important part of contaminating proteins and nucleic acids is denatured and precipitates. These fractions are then centrifuged for 1 hr at 10,000 *g*, and the supernatant is collected.

1.3. Ammonium Sulfate Precipitation

Each liter of the supernatant is dialyzed for 48 hr against 1.5 liters of 3.6-M ammonium sulfate containing 20-mM β-mercaptoethanol. During this step, the preparation is kept at 4°C and the dialysis medium is gently agitated with a magnetic stirrer. The precipitate containing ATCase is collected by centrifugation for 20 min at 20,000 *g* and then redissolved in 70 ml of 10-mM imidazole–HCl buffer (pH 7.0)–2-mM β-mercaptoethanol–0.1-mM EDTA.

1.4. Ion Exchange Chromatography

The solution obtained is dialyzed for 12 hr against 1 liter of the same buffer containing 0.35-M KCl. The preparation is then placed on top of a DEAE–Sephadex A-50 column (6 × 30 cm for about 100 ml of extract) previously equilibrated with 10-mM imidazole–HCl buffer (pH 7.0)–0.5-mM β-mercaptoethanol–0.35-M KCl. When the solution has entirely entered into the gel, the column is eluted with 600 ml of the same buffer, and 10-ml fractions are collected. The ATCase activity of these fractions is determined, using 1-μl aliquots (or, more conveniently, 10 μl of a 10-fold-diluted sample), by one of the methods described in Chapter 5. The active fractions are pooled, and the ATCase solution thus obtained is concentrated by ammonium sulfate precipitation. That is, 1.5 volumes of 3.6-M ammonium sulfate is added, and after 1 hr of stirring at 4°C the precipitate is collected by centrifugation for 20 min at 20,000 *g* and redissolved in 50 ml of 10-mM potassium phosphate buffer (pH 7.0)–2-mM β-mercaptoethanol–0.2-mM EDTA. When using volumes larger than 100 ml, it is more convenient to perform this precipitation by dialysis against the ammonium sulfate solution. Under such conditions, the precipitate can be obtained from a suspension whose volume has been reduced by about a factor of 2 by virtue of the osmotic pressure at the beginning of the dialysis.

1.5. Precipitation at the Isoelectric Point

Fractions (100 ml) of the solution obtained are extensively dialyzed against 2 liters of 100-mM potassium phosphate (pH 5.9)–10-mM β-mercaptoethanol. The buffer, whose pH has to be precisely adjusted, must be renewed several times during the 36–48 hr of this step. After centrifugation for 15 min at 20,000 *g*, the precipitate which contains ATCase and contaminating denatured proteins is resuspended in 20 ml of 10-mM potassium phosphate buffer (pH 7)–2-mM β-mercaptoethanol. The ATCase dissolves, but much of the denatured material will remain as an insoluble residue that can be removed by centrifugation for 15 min at 20,000 *g*. The supernatant is submitted to a second precipitation at pH 5.9. After centrifugation, the pellet is redissolved in 20–30 ml of 40-mM potassium phosphate buffer (pH 7.0)–1-mM β-mercaptoethanol–0.1-mM EDTA. This final solution is clarified by a 30-min centrifugation at 30,000 *g*. The ATCase preparation so obtained can be stored for several years at −20°C in the presence of an equal volume of glycerol.

The purification and yield can be monitored by measuring the ATCase specific activity in the cell extract and in the fractions obtained after each step of the purification. For this purpose, the ATCase activity is measured as described in Chapter 5. The amount of protein is determined as described in Chapter 4. The ATCase specific activity is expressed as micro-moles of carbamylaspartate formed per hour per milligram of protein.

2. DISSOCIATION OF ATCase INTO CATALYTIC AND REGULATORY SUBUNITS

Several methods have been described for dissociating ATCase into isolated catalytic and regulatory subunits (Chapter 1) and for preparing homogeneous solutions of each. A method using a heat shock is very simple and rapid, but it provides only the catalytic subunits. Other methods, based on the reaction of a mercurial reagent (*p*-hydroxymercuribenzoate or neohydrin) with the SH groups of the regulatory subunits, are more elaborate and provide both types of subunits.

2.1. Preparation of Catalytic Subunits by Heat Shock

This method is based on the fact that under particular conditions such as low ionic strength, a heat shock induces the dissociation of

ATCase into the catalytic subunits, which remain active, and regulatory subunits, which are denatured (Gerhart and Pardee, 1963).

Procedure

A sample of ATCase (5–10 mg) is dialyzed for 12 hr against 1 liter of 5-mM potassium phosphate buffer (pH 7.0). The solution obtained is incubated for 5–10 min in a water bath at 60°C. At the end of this incubation, the sample is quickly cooled in ice water and then centrifuged for 15 min at 20,000 *g*. The supernatant contains the isolated catalytic subunits. The success of this procedure can be determined by polyacrylamide gel electrophoresis under nondenaturing conditions by the method described in Chapter 4. A trace amount of denatured regulatory subunits is sometimes visible on the gel, but that does not interfere with the activity of the isolated catalytic subunits.

2.2. Dissociation by *p*-Hydroxymercuribenzoate

In this method, the dissociation of ATCase into catalytic and regulatory subunits results from the reaction of *p*-hydroxymercuribenzoate (pHMB) with the SH groups of cysteine residues that are clustered in the C-terminal region of the regulatory chain. The catalytic and regulatory subunits are then separated by ion exchange chromatography (Gerhart and Holoubek, 1967).

Equipment

- Chromatography column (1.5 × 25 cm)
- Fraction collector
- UV spectrophotometer

Reagents

- pHMB
- DEAE–Sephadex A-50 previously equilibrated with 10-mM imidazole–HCl buffer (pH 7.0)–0.26-M KCl
- Dialysis tubing (1–2 cm in diameter)

Procedure

A 100-mg sample of ATCase in 5–10 ml of buffer is dialyzed for 12 hr against 1 liter of 40-mM potassium phosphate buffer (pH 7.0) to remove the β-mercaptoethanol that is present in the enzyme storage buffer. To this solution is added 15 mg of pHMB dissolved in 0.5 ml of 40-mM Tris–HCl buffer (pH 8.4). The mixture is incubated for 15–20 min at room temperature with gentle agitation. At the end of this incubation, 3-M KCl is added to bring its final concentration to 0.26 M. The sample is then placed on top of a DEAE–Sephadex A-50 column previously equilibrated with the 10-mM imidazole-HCl buffer (pH 7)–0.26-M KCl–0.05-mM pHMB. When the protein solution has entirely settled into the column, it is eluted with the same buffer. Fractions of 2 ml are collected, and their optical density at 280 nm is measured, allowing localization of the peak of regulatory subunits followed by a second peak due to the unreacted excess of pHMB (Fig. 18). The column is then eluted with the same buffer but containing 0.5-M KCl, eluting the catalytic subunits. The fractions collected are kept at 4°C.

Reactivation and Storage of the Subunits

To restore the original properties of the two types of subunits, it is necessary to free the SH groups that have reacted with pHMB as follows.

The fractions containing the ***regulatory subunits*** are pooled, and β-mercaptoethanol and zinc acetate are added to final concentrations of 20 and 2 mM, respectively. The regulatory subunits are then precipitated by dialysis against 50 volumes of 3.6-M ammonium sulfate in 40-mM potassium phosphate buffer (pH 7)–10-mM β-mercaptoethanol–0.1-mM

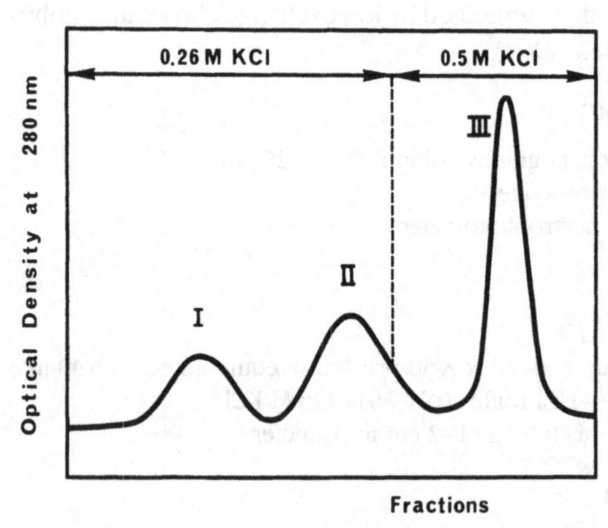

Figure 18. Separation of the two types of subunits of ATCase on DEAE–Sephadex. The experimental conditions are described in the text. I, Regulatory subunits; II, excess pHMB; III, catalytic subunits. (Adapted from Gerhart and Holoubek, 1967, with permission of J. C. Gerhart.)

zinc acetate. The precipitate obtained is collected by centrifugation for 15 min at 20,000 *g* and redissolved in about 5 ml of the same buffer but without ammonium sulfate. To eliminate the remaining traces of this salt, the preparation is dialyzed for 12 hr against the same buffer.

The fractions containing the *catalytic subunits* are pooled, and β-mercaptoethanol is added to a final concentration of 10 mM. This solution is concentrated by dialysis against ammonium sulfate as described above for the regulatory subunits except that zinc acetate is replaced by 0.2-mM EDTA. The precipitate obtained is redissolved in 5–10 ml of 40-mM potassium phosphate buffer (pH 8.0)–1-mM β-mercaptoethanol–0.1-mM EDTA.

The catalytic and regulatory subunits are kept at 4°C or, for long-term storage, at −20°C. The protein concentration of these solutions is determined as described in Chapter 4. A rapid estimate of the catalytic subunit concentration can be obtained on the basis of the absorbance of the solution at 280 nm, using the absorption coefficient given in Table 1. With respect to the regulatory subunits, this kind of measurement is more questionable since the solution is often contaminated by traces of pHMB, which absorbs significantly at 280 nm.

2.3. Dissociation by Neohydrin

The principle of this method is the same as in the case of pHMB. Its advantage lies in the fact that it is faster and provides a better yield for the regulatory subunits (Cohlberg *et al.*, 1972).

Equipment

- Chromatography column (0.5 × 2 cm)
- Fraction collector
- UV spectrophotometer

Products

- Neohydrin [1- (3-chloromercuri-2-methoxypropyl)-urea]
- DEAE–cellulose

Procedure

A 100-mg sample of ATCase in 5–10 ml of buffer is dialyzed for 12 hr against 1 liter of 10-mM Tris–HCl (pH 8.3)–0.1-M KCl to eliminate the β-mercaptoethanol present in the storage buffer. To the solution obtained, 10 mg of neohydrin in 1 ml of Tris–HCl buffer (pH 9.3) is added. After 15 min of incubation at room temperature, the sample is placed on top of a DEAE–cellulose column previously equilibrated with 10-mM Tris–HCl buffer (pH 8.3)–0.1-M KCl–0.05-mM neohydrin. When all of the sample has entered the column, elution is performed with the same buffer at a rate of 25 ml/hr. Fractions of 2 ml are collected, and their optical density at 280 nm is measured. After the appearance of the peak of regulatory subunits followed by unreacted neohydrin, the elution is continued, using the same buffer containing 0.5-mM KCl to release the catalytic subunits.

The fractions containing the two types of subunits are pooled and treated as described for the preceding method.

3. RECONSTITUTION OF ATCase FROM ITS ISOLATED CATALYTIC AND REGULATORY SUBUNITS

Procedure

To a 100-μl solution of regulatory subunits at 2.5 mg/ml, add 200 μl of 40-mM potassium phosphate buffer (pH 7.0)–2-mM β-mercaptoethanol–0.2-mM zinc acetate. To this mixture add 100 μl of catalytic subunit solution at 2.5 mg/ml, and incubate the mixture for 30 min at 37°C. The result is analyzed by polyacrylamide gel electrophoresis under nondenaturing conditions as described in Chapter 4. The same preparation is also used to determine enzymatic activity (Chapter 5).

The stoichiometry of reassociation of the subunits can be investigated under the same conditions by varying the amounts of regulatory subunits that are added to the catalytic subunits. For this purpose, increasing volumes of the regulatory subunit solution from 25 to 100 μl should be added to the catalytic subunits, the differences in volume being compensated for by complementary volumes of the buffer used to store the subunits.

Analyze the results of these incubations by polyacrylamide gel electrophoresis as previously described.

Chapter 4

Structural and Physicochemical Study of Aspartate Transcarbamylase

Several methods allow us to estimate the purity of the ATCase preparations and to obtain some structural information about this enzyme. Combination of the results obtained by using different techniques allows the determination of the stoichiometry of assembly of the catalytic and regulatory chains.

1. ABSORPTION SPECTROPHOTOMETRY AND DETERMINATON OF PROTEIN CONCENTRATION

1.1. Theory

The visible and UV absorption spectra of macromolecules in solution cannot provide direct information about their three-dimensional structure (Tanford, 1961). However, physicochemical studies of these molecules often begin by determining their electronic spectra in solution, which can provide useful information regarding their composition, purity, concentration and in some cases (indirectly) their structure. In the case of proteins, UV absorption (between 250 and 300 nm) is almost entirely due to the indole group of tryptophan residues, to the phenol group of tyrosines, and to a slight contribution of cysteines and the phenyl group of phenylalanines. The absorption spectra of these different chromophores are shown in Fig. 19. A typical protein solution spectrum is shown in Fig. 20. For a 1-mg/ml solution, the maximal

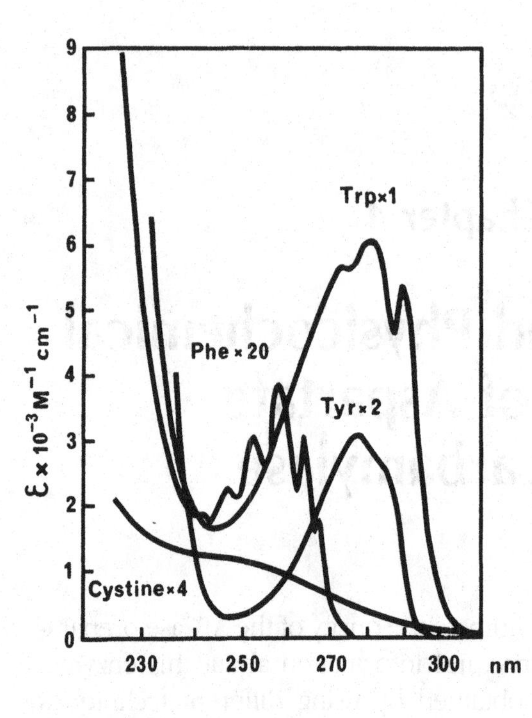

Figure 19. Absorption spectra of the amino acids that contribute to the spectrum of a protein. These spectra were obtained in methanolic solutions, using the *N*-acetyl ethyl ester derivatives of phenylalanine, tyrosine, and tryptophan and the dimethyl ester of cystine. To facilitate comparison, these spectra are enlarged by the factors indicated. ε is the Molar extinction coefficient. (Adapted from Metzler *et al.*, 1972, with permission of D. E. Metzler.)

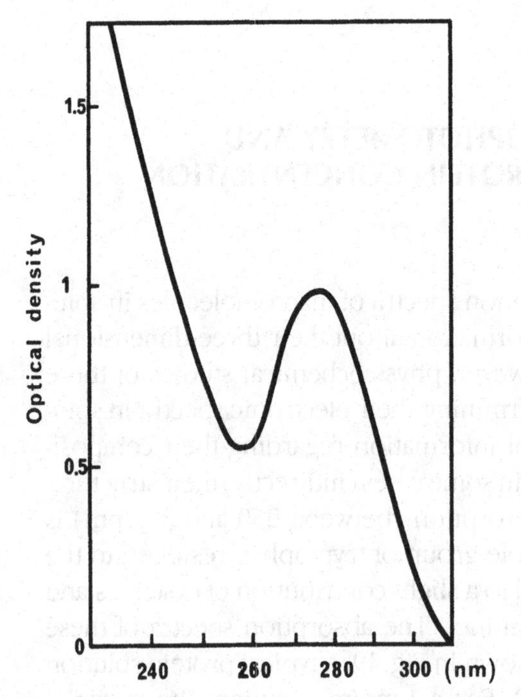

Figure 20. Typical absorption spectrum of a protein. This spectrum is that of a 1-mg/ml protein solution at pH 7 for an optical path of 1 cm.

absorption, centered at around 280 nm, varies from 0.5 to 1.5, depending on the content of the protein in chromophores. It can be seen on the same figure that the ratio of absorption at 280 and 260 nm is about 1.75. The optical density at 235 nm, essentially due to the peptide bonds, is higher than that at 280 nm.

The absorption spectrum of a protein is very close to the theoretical one, which can be calculated as the sum of the absorptions of its constituent amino acids. For instance, in Fig. 21, comparison of the solid and dashed lines shows that the absorption of albumin is very close to that of the amino acid mixture obtained after hydrolysis. On the basis of the amino acid absorption spectra (Fig. 19), the tryptophan and tyrosine content of most proteins can be determined from their absorption at two wavelengths at alkaline pH (Bencze and Schmid, 1957). The influ-

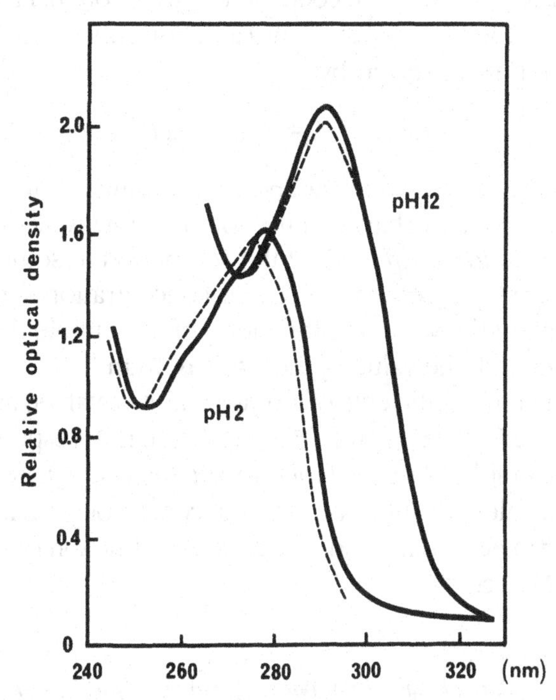

Figure 21. Influence of pH on the absorption spectra of albumin and the product of its hydrolytic degradation into amino acids. Dashed lines indicate the spectra of native albumin; solid lines indicate the corresponding mixture of free amino acids obtained by hydrolysis. At alkaline pH, phenol groups are present as phenolate. (Adapted from Beaven and Holiday, 1952, with permission.)

ence of pH on absorption spectra (Fig. 21) results from the fact that some amino acids, such as tyrosine, bear ionizable groups. In addition, the spectra are influenced by the ionic or the nonpolar environment of the chromophores.

Determination of the *exact concentration* of protein solutions is important. Their UV absorbance or their absorbance after reaction with a chromogenic reagent (biuret method; methods of Lowry *et al.* and Bradford; etc.) provides only a *relative determination* of its concentration in comparing it with a solution of a reference protein whose exact concentration is known. In general, bovine serum albumin is used for that purpose. The exact concentration of a protein solution can be determined either on the basis of its dry weight or by quantitative amino acid analysis. When the amino acid composition of the protein is known, another way is to measure the nitrogen content of its solution. On the basis of the exact concentration and optical density of a protein solution, one can calculate the extinction coefficient of the protein. From the law of Beer-Lambert, the optical density of a solution submitted to a monochromatic light beam is given by:

$$\log I_0/I = A = OD = \varepsilon c l \quad (1)$$

in which I_0 and I are the incident and the transmitted light intensity, respectively, A is the absorbance of the solution, OD is its optical density, ε is the *extinction coefficient* of the compound in solution (milliliters per milligram per centimeter), c is its concentration in milligrams per milliliter, and l the width (centimeters) of the sample that absorbs the light. In general, the value of the latter is 1 cm.

The extinction coefficients of ATCase and its catalytic and regulatory subunits are 0.59, 0.72, and 0.35 ml mg^{-1} cm^{-1}, respectively. It is sometimes preferable to use the *molar extinction coefficient*, especially for small molecules. In this case, the concentration is expressed as moles per liter in equation 1 and ε (the molar extinction coefficient) is expressed as M^{-1} cm^{-1}.

Question 3. *On the basis of the values given in Table 1, calculate the molar extinction coefficients of ATCase and its catalytic and regulatory subunits.*

1.2. Experimental Determinations: Assay of Protein Concentration

The relative concentrations of the solutions of ATCase or its isolated subunits can be determined either by a direct measurement of UV absorption or by chemical assays using a solution of bovine serum albumin as the standard.

1.2.1. Spectrophotometric Method

There is a purely empirical spectrophotometric method that enables one to estimate approximately the concentration of a protein solution (Whitaker and Granum, 1980). In this simple and rapid method, the optical density of the solution is measured at 235 nm (peptide bond absorption) and at 280 nm (aromatic side chain absorption) for an optical path of 1 cm. Under these conditions, the protein concentration (in milligrams per milliliter) is given by

$$(OD_{235} - OD_{280})/2.51$$

This method does not require the use of a reference protein solution. The concentration of the sample must be in the range of 0.5–1.5 mg/ml. In this method, the reference cuvette of the spectrophotometer must contain the buffer in which the protein is dissolved.

Results

Compare the values obtained with those calculated from the optical density at 280 nm and the absorption coefficients given in Table 1.

1.2.2. Chemical Methods

Several colorimetric assays enable the concentrations of protein solutions to be measured.

1.2.2a. Method of Lowry et al. (1951). This method is based on the biuret reaction and on the use of the Folin–Ciocalteu reagent, which reacts specifically with the phenol group of tyrosine. Consequently, the result obtained with a given protein will depend on its content in this amino acid.

Equipment

- Spectrophotometer for visible light

Reagents

- α: Sodium carbonate at 20% in 0.1-N sodium hydroxide
- β: Copper sulfate at 1%
- γ: Sodium–potassium bitartrate at 2%
- δ: Folin–Ciocalteu reagent

Procedure

The active reagent made of 1% β and 1% γ in α is prepared just before use. To samples containing 5–75 μg of protein in 0.2 ml, add 1 ml of this reagent. Each series of samples is compared to a reference curve obtained by using a solution of 1 mg/ml bovine serum albumin as the standard. For an ATCase solution of about 1 mg/ml, adequate conditions are as in Table 3.

To each of these samples, add 1 ml of the active reagent. After shaking and incubating the mixture for 10 min at room temperature, add it to 0.1 ml of a 1:1 mix of Folin–Ciocalteu solution and distilled water. After agitation, incubate this mixture for 20 min at room temperature. Using a spectrophotometer, measure the optical density at 750 nm, using sample 1, which contains only the reagents and buffer, as a blank. Draw the standard curve of the optical density as a function of serum albumin concentration. Use this curve to determine the protein concentration of the ATCase solutions.

As previously mentioned, the color intensity depends on the tyrosine content of the protein being analyzed. With respect to ATCase, the

Table 3. Determination of ATCase Concentration by the Method of Lowry *et al.* (1951)

	Sample Number									
Volume (μl)[a]	1	2	3	4	5	6	7	8	9	10
Serum albumin		10	25	50	75	100				
ATCase							10	25	40	60
ATCase conservation buffer	60	60	60	60	60	60	50	35	20	
Distilled water	140	130	115	90	65	40	140	140	140	140

[a]The final volume of each sample is 200 μl.

Lowry assay performed with a serum albumin standard curve leads to a 20% overestimation of the concentration of this enzyme (Kerbiriou *et al.*, 1977).

Question 4. *What remarks can be made concerning the bovine serum albumin standard curve?*

1.2.2b. Bio-Rad Colorimetric Assay. The dye used in this method reacts with proteins to give a colored derivative whose concentration is measured on the basis of its optical density at 595 nm. As with the Lowry assay, the measurements are made in comparison with a standard curve of bovine serum albumin prepared in distilled water.

Equipment

- Spectrophotometer for visible light

Reagent

- Bio-Rad reagent

Procedure

One volume of Bio-Rad reagent is added to 4 volumes of the protein solution (1–50 μg). After agitation and a 5-min incubation at room temperature, the optical density at 595 nm is measured within 2 hr after the addition. For an ATCase solution of about 1 mg/ml, use the conditions given in Table 4.

Never mouth pipette the Bio-Rad reagent, which contains some perchloric acid and methanol.

Numerous chemicals interfere with these colorimetric assays [sodium dodecyl sulfate (SDS), salts, SH groups, sugars, etc.], generally increasing the coloration obtained. For this reason, it is important to add identical volumes of conservation buffer to all samples and standards.

2. ANALYSIS BY ELECTROPHORESIS

2.1. Theoretical Aspects

The analysis of biological marcomolecules largely involves the use of electrophoretic techniques.

Table 4. Determination of ATCase Concentration by Using Bio-Rad Reagent

Volume (μl)[a]	Sample number 1	2	3	4	5	6	7	8	9	10	11
Serum albumin, 1 mg/ml					5	10	15	20	30	40	50
ATCase		15	25	50							
ATCase conservation buffer	100	85	75	50	100	100	100	100	100	100	100
Distilled water	1500	1500	1500	1500	1495	1490	1485	1480	1470	1460	1450
Bio-Rad reagent	400	400	400	400	400	400	400	400	400	400	400

[a]The total volume of each sample is 2 ml.

2.1.1. The Moving-Boundary Method

2.1.1a. General Case. A single particle bearing a charge Q (coulombs) placed in an electrical field $\mathbf{E}$ (volts per centimeter) is subjected to an electrical force $\mathbf{F} = Q\mathbf{E}$ (dynes) which is oriented in the direction of the field. Such a particle in a viscous fluid is also subjected to a frictional force which is oriented in the opposite direction. This frictional force is proportional to the speed of the particle provided that it is not too high. That is, $-\mathbf{F}' = f\mathbf{v}$, where $\mathbf{v}$ is the speed of the particle in centimeters per second, and f is the frictional coefficient. If the particle is a sphere of radius R (or R_s; see Section 4.2), one has:

$$f = 6\pi\eta R \quad (2)$$

where η is the viscosity (in poise) of the medium.

The particle will accelerate until an equilibrium is established between these two opposite forces, i.e., $\mathbf{F} - \mathbf{F}' = 0$. The rate of migration is then constant and

$$\mathbf{v} = Q\mathbf{E}/f = Q\mathbf{E}/6\pi\eta R \quad (3)$$

The electrophoretic mobility u of a particle is, by definition, its rate of migration in a homogeneous field equal to unity ($\mathbf{E} = 1$ V/cm). Thus, for a given viscosity, the electrophoretic mobility depends only on the molecular parameters of the particle: its charge and radius.

$$u = Q/6\pi\eta R \quad (4)$$

2.1.1b. Macro-Ions. Proteins and nucleic acids in solution are charged as the result of the ionization of some carboxyl, amine, or phosphate groups. This ionization depends on a large number of factors but results mainly from the pH and ionic strength of the medium. For example, it is easy to understand that the net charge of most proteins dissolved in an acidic medium will be positive. If two electrodes are placed in this medium and if a tension V (in volts) is applied, an electrical field will be generated in the solution; the protein will behave as a cation and migrate toward the cathode at a speed v.

In the case of macro-ions, the theory is much more complex than in the case of a singly charged particle. The rate of migration will depend not only on the charge but also on the layer of counterions present in their vicinity. The counterions are small molecules always present in solution (e.g., Na^+, Cl^-, K^+, or HCO^{3-}) with charges opposite in sign to those of the macro-ion. The electrical field that acts on the macro-ion also acts on the counterions but in the *opposite direction*. The counterions carry along some solvent molecules, thus slowing down the migration of the macro-ions. Under these conditions, equation 4 becomes

$$u = (Q/6\pi\eta R)(1/Y) \quad (5)$$

where Y is a complex function that takes into account the presence of the counterions, the ionic strength of the medium, etc. (Tanford, 1961). Despite elaborate theoretical developments, no accurate relationship between electrophoretic mobility and molecular parameters is yet available in the case of macro-ions. However, some very complex equations can be used to provide *semiquantitative* information on the molecular charge, the size, and the shape of proteins. In practice, the moving-boundary method of electrophoresis is now very rarely used in laboratories.

2.1.2. Gel Electrophoresis

Electrophoresis has become essentially a method used for the separation of macromolecules and under some conditions for the determination of their apparent molecular mass. One uses matrices such as agarose or polyacrylamide gel which facilitate the experiments and eliminate convection currents.

2.1.2a. Electrophoresis under Nondenaturing Conditions. Polyacrylamide gels are obtained by the polymerization of acrylamide and a reticulating agent (*bis*-acrylamide) in a buffer that contains a catalyst. The latter can be riboflavin, ammonium persulfate, N,N,N',N'-tetramethylethylenediamine (TEMED), or any other agent able to produce free radicals. The reticulated gel has the following structure:

```
 H                              H
 |                              |
 CH—CO—NH2                      CH—CO—NH2
 |          a                   |          c
 CH2                            CH2
                                |
 |                              |
 CH—CO—NH—CH2—NH—CO—CH
 |                              |
 CH2                            CH2
 |                              |
 CH—CO—NH2                      CH—CO—NH2
 |          b                   |          d
 CH2                            CH2
 |                              |
```

in which a, b, c, and d are variable. The degree of reticulation depends on the proportion of acrylamide and *bis*-acrylamide. In general, gels made from 7.5% acrylamide and 0.2% *bis*-acrylamide are used.

Proteins subjected to an electrical field move proportionally to their charge and as an inverse function of their molecular size (Eq. 5). When they migrate within a reticulated gel, their size is involved in two ways:

1. As in the case of the moving-boundary method of electrophoresis, the friction force is proportional to the radius R of the molecule.
2. Due to the reticulation of the gel, products are retarded in proportion to their size.

At first approximation, equation 5 becomes:

$$u = (Q/6\pi\eta R)(1/Y)(1/K) \qquad (6)$$

where K is a factor which depends on both the size and the shape of the protein and takes into account the effect of the reticulation.

Thus, the electrophoretic mobility of a protein in a gel depends not only on its molecular mass but also on other parameters that are difficult to distinguish in practice. Nevertheless, gel electrophoresis under nondenaturing conditions is often used to separate and visualize different molecular species. It will be used here to control the purity of ATCase preparations and to estimate the efficiency of the separation and recombination of its catalytic and regulatory subunits.

2.1.2b. Electrophoresis under Denaturing Conditions. In this case, the gel is prepared in the same way as previously described but using a buffer that contains the anionic detergent SDS. The proteins are dissolved in the presence of a high amount of SDS and β-mercaptoethanol, which reduces the disulfide bonds. The mixture is heated either for a few minutes at 100°C or for a longer period of time at a lower temperature. Under these conditions, it was shown (Reynolds and Tanford, 1970) that generally:

1. Proteins are dissociated into their constituent polypeptide chains; for instance, a dimeric protein is dissociated into monomers.
2. They bind a fixed amount of SDS (about 1.4 g per gram of protein).
3. Their structure changes (denaturation), and they behave as elongated cylinders like fibrinogen or myosin. However some SDS–protein complexes can be more flexible, especially long polypeptide chains at high ionic strength (Makino, 1979).

Under these conditions, the mobility of proteins on SDS–polyacrylamide gels depends not so much on their charge but on their molecular mass. The intrinsic charge of the protein is obliterated by that of the bound molecules of SDS. For instance, ovalbumin ($M_r = 43{,}000$) possesses about ten net negative charges at pH 7; the addition of an excess of SDS increases its number of negative charges to about 200. Under these conditions, all proteins have an elongated shape and bear about the same number of negative charges per unit of length. Thus, at first approximation the ratio Q/R in equation 6 is constant. Similarly, factor Y will increase linearly with the molecular mass. Consequently, the mobility of the proteins will vary as a function of Y and will be subjected to the sieving effect of the gel (factor K in equation 6). The distance of migration after a given time will be inversely proportional to the molecular mass.

The experimental data show that there is a linear relationship between the electrophoretic mobility and the logarithm of the protein molecular mass (see, for example, Weber and Osborn, 1969). This is true only within a limited range of molecular mass (from about 10,000–100,000 daltons for a 7.5% acrylamide–0.2% *bis*-acrylamide gel). Using proteins of known molecular mass as standards, it is possible to calibrate a gel and to determine the apparent molecular mass of an unknown protein. However, when one is using this method it is very important to bear in mind the following causes of error:

1. The reference proteins are water-soluble globular proteins. It would be hazardous to use this method to determine the molecular mass of other types of proteins (membrane proteins, glycoproteins, etc.). They do not necessarily bind 1.4 g of SDS per gram (Rizzolo *et al.*, 1976); moreover, they can exhibit hydrodynamic behavior different from that of the reference proteins.
2. The intrinsic charge of the protein is not taken into account, and this approximation is not always valid. For instance, the molecular masses of very basic proteins such as ribosomal proteins or histones are often overestimated when determined by this method.
3. Some proteins are not dissociated into their polypeptide constituents or are not entirely denatured in the presence of high concentrations of SDS. For instance, calmodulin is still able to bind calcium in the presence of 0.1% SDS (Cox and Stein, 1981).

Thus, this empirical technique must be used with caution. It provides only an *apparent molecular mass* by reference to standard proteins. Direct measurements of *absolute molecular mass* can be performed by using other methods such as sedimentation equilibrium, which will be described later.

SDS–gel electrophoresis is used here to estimate the apparent molecular mass of the polypeptide chains that constitute the catalytic and regulatory subunits of ATCase.

2.2. Experimental Determinations

Two types of apparatus can be used for gel electrophoresis. The first makes use of cylindrical gels in small glass tubes, each one receiving a single sample. The second, now more common, makes use of thin, flat gels that are prepared between two glass plates (**slab gels**). Such a gel can accept about ten samples, which migrate in parallel.

In a practical involving a large group of students, the use of slab gels may be more difficult. In addition, under nondenaturing conditions the cylindrical gels facilitate the identification of ATCase in bacterial extracts. The advantages of the slab gels are discussed later.

2.2.1. Electrophoresis in Cylindrical Gels

The technique makes use of an Acrylophore-type apparatus, which consists of an upper and a lower reservoir, containing the same buffer, in which a circular platinum electrode is immersed (Fig. 22). Small glass tubes containing the gels are held vertically between the two reservoirs by ring gaskets, which ensure watertightness. A stabilized direct current power supply is connected to the electrodes. The polarity is chosen to ensure the migration of the molecular species towards the lower electrode, taking into account their net electric charge.

2.2.1a. Electrophoresis under Nondenaturing Conditions. Each tube (internal diameter, 0.6 cm; length 7 cm) contains two different gels:

1. An upper gel (2.5% acrylamide in Tris–H_3PO_4 buffer, ph 6.9) whose degree of reticulation is low, allowing fast migration of all proteins; this phenomenon leads to the concentration of all proteins contained by the sample into a thin layer.
2. A lower gel (7.5% acrylamide in Tris–glycine buffer, pH 8.3) with a high degree of reticulation, which allows separation of the various proteins during migration.

Solutions

The following solutions must be kept at 4°C. Never mouth pipette solutions C and D.

• Solution A, pH 9.2	1-N HCl: 48 ml Tris: 36.3 g TEMED: 0.46 ml H_2O: up to 100 ml
• Solution B, pH 6.9	1-M H_3PO_4: 25.6 ml Tris: 5.7 g H_2O: up to 100 ml

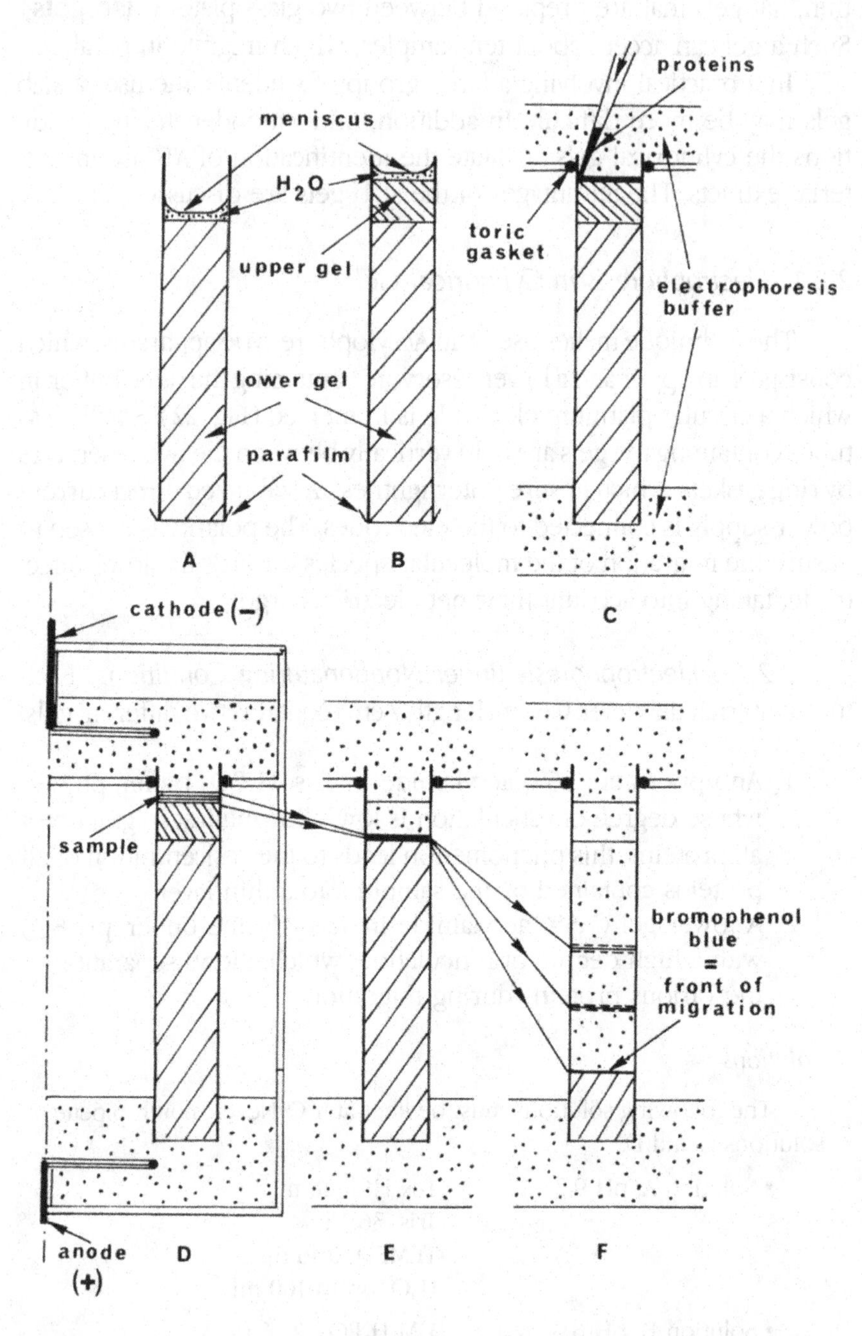

Figure 22. Electrophoresis under nondenaturing conditions; preparation and use of the cylindrical gels.

• Solution C, stored in a dark bottle	Acrylamide: 30 g *Bis*-acrylamide: 0.8 g H_2O: up to 100 ml
• Solution D, stored in a dark bottle	Acrylamide: 10 g *Bis*-acrylamide: 0.29 g H_2O: up to 100 ml
• Solution E, stored in a dark bottle	Riboflavin: 4 mg H_2O: up to 100 ml
• Solution F prepared just before use	Ammonium persulfate: 200 mg H_2O: up to 100 ml
• Tris–glycine buffer, pH 8.3	Tris: 6 g Glycine: 26.8 g H_2O: up to 1000 ml
• 3 (dimethylamino)propionitrile (DMAP) at 10%	
• Sample loading solution, containing the indicator	Glycerol: 50% Bromophenol blue: 0.01%
• Staining solution	Coomassie blue G250: 0.04% in 3.5% perchloric acid

Gel Preparation

- Seal the lower end of the electrophoresis tubes with a small piece of parafilm (Fig. 22A).
- The lower gel is prepared from the following mixture (for eight gels, 1.5 ml each):

  2 ml of A
  4 ml of C
  2 ml of H_2O
  8 ml of F
- Fill the tubes with this mixture up to 1.5 cm from the top with a Pasteur pipette.
- Still using a Pasteur pipette, carefully deposit a small volume of water on top of this lower gel to obtain a very flat water–gel interface (Fig. 22A). This lower gel polymerizes within about 30 min. The water–gel interface then becomes very apparent.
- Remove the water layer with absorbing paper.
- The upper gel is then prepared from the following mixture (for eight gels, 0.4 ml each):

  0.5 ml of B
  1 ml of D

0.5 ml of E
2 ml of H_2O
10 μl of 10% DMAP

- Deposit about 0.4 ml of this mixture on top of the lower gel to obtain an upper gel about 1 cm high (Fig. 22B).
- Carefully layer a small amount of water on top of this gel.
- Polymerize the gels by exposure for about 30 min to the light of a bright lamp. The gels thus obtained can be stored several days at 4°C.
- Position the tubes on the apparatus.
- Remove the parafilm stopper.
- Fill the two reservoirs with the Tris–glycine buffer (pH 8.3) to an adequate level, avoiding the formation of air bubbles at both ends of the gels.

Samples

Each protein sample is diluted by an equal volume of the 50% glycerol solution of bromophenol blue (maximum final volume; 100 μl). The concentrated glycerol increases the density of the sample and facilitates its layering. The bromophenol blue will be the fastest-migrating species and will define the front of migration needed to calculate the *Rf* (see Results).

After mixing (vortex), deposit the samples on top of the gels, using a micropipette whose end is placed very close to the surface of the gel (Fig. 22C). Note the number of the tube and the nature of the sample.

Migration

In the buffer used (Tris–glycine, pH 8.3), the majority of the proteins migrate toward the positive pole; it is therefore necessary to connect the anode to the lower part of the apparatus. Adjust the current intensity to about 7 mA per tube. The migration (Fig. 22D–F) requires about 1 hr. Stop the migration when the bromophenol blue band reaches the bottom of the gel.

Staining

Cautiously remove the gels from the tubes: using a syringe equipped with a fine needle, inject water between the tube and the gel all around the circumference of the gel to liberate it from the tube. It is often necessary to do this at both ends of the gels.

Cut the lower gel with a razor blade in the middle of the bromophenol blue band, which will disappear during fixation, and put the gels in numbered tubes containing the staining solution (0.04% Coomassie blue G250 in 3.5% perchloric acid).

Protein bands begin to appear after 10 min but it is recommended that the coloration be continued for about 12 hr to obtain a more intense staining, allowing the detection of minor bands. Remove the dye and replace it with distilled water. The gels may be stored in 5% acetic acid.

Electrophoretic Analysis of ATCase and Its Isolated Subunits

Use 20-μl samples of 0.5-mg/ml solutions of ATCase, catalytic subunits, regulatory subunits, and reconstituted enzyme (see Chapter 3). To those samples add 20 μl of the sample loading solution containing the indicator, and deposit the mixture on top of the gels as previously described.

Results

- Note the experimental conditions used (intensity, time of migration, etc.).
- Carefully record the position of the protein bands obtained, by drawing the gels real size on graph paper.
- Calculate the *Rf* of each protein band:

  Rf = distance of migration of the protein/distance of migration of the indicator.

> Question 5. *What are the appearance, the number, and the relative proportions of the protein bands obtained for each of the samples analyzed? What conclusions can be drawn?*

Identification of ATCase in Bacterial Extracts by Electrophoresis

Polyacrylamide gel electrophoresis under nondenaturing conditions allows one to identify the protein band in the crude bacterial extract that is responsible for the ATCase activity. For this purpose, electrophoresis is carried out simultaneously on two identical gels. At the end of migration, one of these gels is stained as previously described and constitutes the reference. The second one is immediately frozen on a dry-ice plate to avoid diffusion of the protein bands. After partial thawing, this gel is cut laterally into 1-mm-thick slices. This can be done either with an automatic slicer or more simply by hand with a razor blade. Each slice is suspended in 250 μl of the ATCase storage buffer. To each of these suspensions, add 250 μl of the incubation medium used for the measurement of ATCase activity described in Chapter 5, which contains the incubation buffer and the two substrates of the enzyme. Incubate the suspensions at 37°C for 30–60 min before determining the amount of carbamylaspartate formed. Compare the result obtained with the reference gel.

2.2.1b. Electrophoresis under Denaturing Conditions. In this method (Weber and Osborn, 1969), one uses a single type of gel, 10% acrylamide in phosphate buffer (pH 7), containing SDS to denature the proteins.

Solutions (store at 4°C, except for Solutions I and IV).

- Solution I
  $NaH_2PO_4 \cdot 2H_2O$: 8.8 g
  $Na_2HPO_4 \cdot 12H_2O$: 51.6 g
  SDS: 2 g
  H_2O: up to 1000 ml

This solution must be kept at room temperature, since SDS precipitates at low temperatures. In addition, avoid the presence of K^+ ions, since potassium dodecyl sulfate is not very soluble.

- Solution II
  Acrylamide: 22.2 G
  Bis-acrylamide: 0.6 g
  H_2O: up to 100 ml

Store this solution in a dark flask; do not mouth pipette.

- Solution III
  Ammonium persulfate: 1.5 g
  H_2O: 100 ml
- Solution IV
  10 mM sodium phosphate buffer, pH 7.0
  SDS: 2%
  β-mercaptoethanol: 2%

This solution must be kept at room temperature.

- Sample loading solution, containing the indicator
  Glycerol: 50%
  Bromophenol blue: 0.01%
- TEMED
- SDS staining solution
  Coomassie blue R250: 1.25 g
  Methanol: 227 ml
  H_2O: 227 ml
  Acetic acid at 100%: 46 ml

- Destaining solution — Methanol: 50 ml
  H_2O: 875 ml
  Acetic acid at 100%: 75 ml

Gel Preparation

The gels are prepared as described above for electrophoresis under nondenaturing condition but using the following solutions (for eight gels, 2 ml each):

Solution I: 10 ml
Solution II: 9 ml
Solution III: 1 ml
TEMED: 30 μl

In this case, a single type of gel is used. The tubes are filled up to 0.5 cm of the upper end. Do not forget to deposit a small layer of water at the surface of the polymerizing gel in order to obtain a flat surface on top. Fill the two reservoirs of the apparatus with solution I diluted with an equal volume of water.

Molecular Mass Markers

Since electrophoresis under denaturing conditions allows an estimation of the molecular mass of the proteins being studied, one uses as standards a series of proteins whose molecular masses are known. For instance:

	(M_r)
Cytochrome *c*	12,400
Myoglobin	16,900
β-Lactoglobulin	18,400
α-Chymotrypsinogen	25,700
Ovalbumin	43,000
Bovine serum albumin	66,000

In the native state, all of these reference proteins are monomers except β-lactoglobulin, which is a dimer of molecular mass 36,800. Stock solutions of each of these proteins are prepared at 1 mg/ml in 10-mM sodium phosphate buffer, pH 7.

Treatment of Markers and Samples by SDS

Before being deposited on top of the gels, the samples must be treated to denature the proteins. This is done in the following way:

- 10 to 15-μl samples of each marker solution and of the sample to be analyzed are mixed to 20 μl of solution IV in numbered small test tubes able to resist exposure at 100°C.

- Incubate these samples for at least 1 min in boiling water.
- Add 30 μl of the sample loading solution containing the indicator and mix. After stirring, the samples are ready. Place the whole of these samples on top of the gels.

Migration

Run the electrophoresis with an intensity of 2 mA per tube; the proteins migrate toward the anode. Under these conditions, the migration requires about 15 hr but does not cause too much heating by the Joule effect.

Staining and Destaining

Carefully remove the gels from the tubes as previously described. Using a razor blade, cut the gels at the level of the bromophenol blue band. Put the gels in numbered test tubes containing the SDS staining solution, noting their orientation. After at least 30 min of staining, remove the dye solution and replace it by the destaining solution, which must be renewed five times within 24 hr. The destaining process can be accelerated by placing the sealed tubes in a 40°C water bath for 1 hr. The proteins appear blue. The gels can be stored in 5% acetic acid.

Analysis of ATCase and Its Subunits

Proceed as described above, using 1-mg/ml solutions of ATCase, catalytic subunits, and regulatory subunits.

Results

- Record the positions of the protein bands by drawing the gels real size on graph paper.
- Calculate the Rf of each protein band.
- Draw the curve of $\log Mr = f(Rf)$ for the standard proteins used. Function f takes into account the different parameters of equation 6, and it can be approximated to a constant. Make use of the curve obtained to estimate the molecular mass of the catalytic and regulatory chains of ATCase.

2.2.2. Slab Gel Electrophoresis

Slab gels are now more commonly used than cylindrical gels. The gel composition is the same as that previously described, but the polymerizing mixture is poured between two parallel glass plates separated by about 1 mm. Placing a comb at the top of the gel before polymerization ensures the formation of wells in which the samples can be deposited.

For electrophoresis under denaturing conditions, the method of Laemmli (1970) allows a higher resolution than does the simplified method previously described.

Solutions

• Solution a, pH 8.8	Tris: 45.4 g
• Solution b, pH 6.8	Tris: 3 g SDS: 0.2 g H_2O: up to 50 ml, (pH adjusted with 37% HCl)
• Solution c (filtered on Millipore HAWP, 0.45 μm)	Acrylamide: 30 g *Bis*-acrylamide: 0.8 g H_2O: up to 100 ml
• Solution d, pH 8.8	Tris: 6 g Glycine: 28.8 g SDS: 2 g H_2O: up to 2000 ml
• TEMED	
• Solution e, prepared just before use	Ammonium persulfate: 10 g H_2O: 100 ml
• Solution f, pH 6.8	Tris: 12 g SDS: 4 g β-Mercaptoethanol: 2 ml 100% glycerol: 20 ml H_2O: up to 100 ml, (pH adjusted with 37% HCl)
• Staining solution (filtered)	Coomassie blue R250: 0.2 g Methanol: 45 ml 100% acetic acid: 10 ml H_2O: up to 100 ml
• Destaining solution	Methanol: 20 ml 100% acetic acid: 10 ml 100% glycerol: 1 ml H_2O: up to 100 ml

- Solution for drying — 100% glycerol: 6 ml; 100% acetic acid: 4 ml; H_2O: up to 100 ml
- Indicator solution — Bromophenol blue: 0.1%, in 20-mM Tris-HCl, pH 7

Gel Preparation

Two square glass plates (side = 16.5 cm), one of which has ears formed by cutting out a section measuring 13.5 × 1 cm, are assembled as shown in Fig. 23. These plates must be perfectly clean. Spacers 0.8 mm

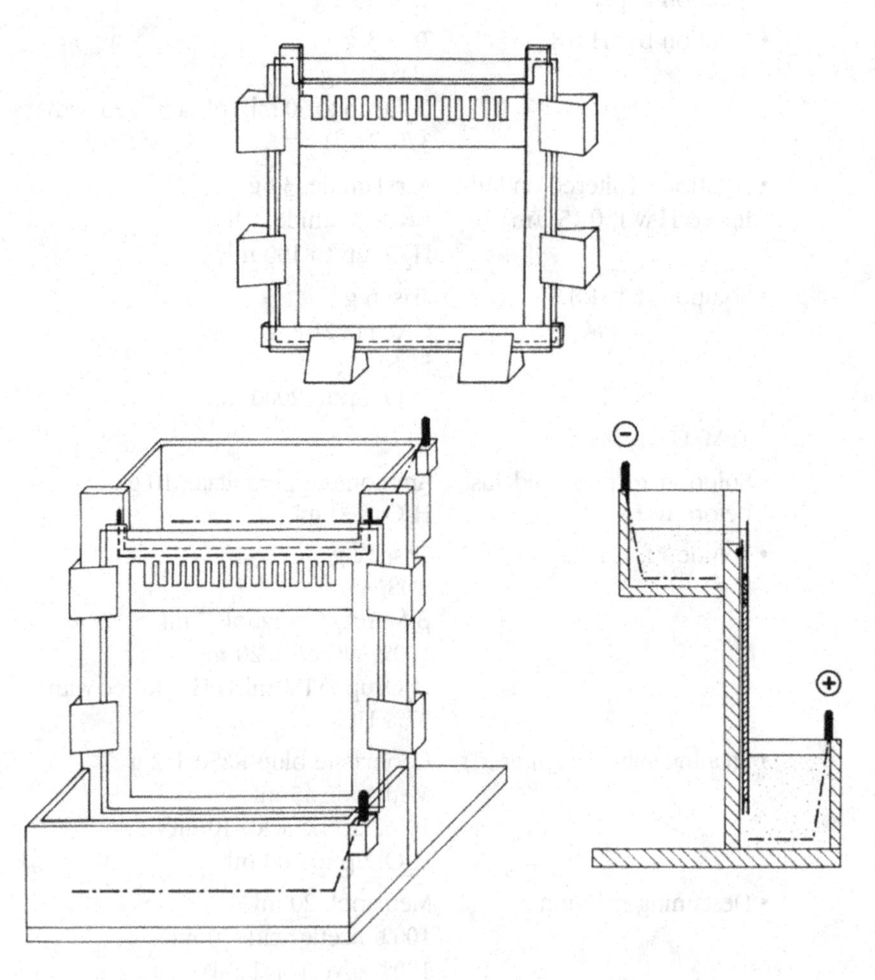

Figure 23. Design of a vertical electrophoresis unit for slab gels.

thick are placed 2–3 mm from the three straight sides of the plates, which are held together by clamps. The plates are sealed together by pouring a liquid agarose solution outside the spacers. This agarose solution is prepared by dissolving 1 g in 100 ml of boiling water and is cooled to about 45°C before use. Solidification of the agarose must ensure complete watertightness of the casting stand. This will be easier to achieve if the surface of the spacer is rugged. The acrylamide gel must then be poured between the two plates within 2 hr after the agarose sealing.

To obtain the chosen acrylamide concentration the lower gel must have the following composition:

	7.5%	10.4%	12.5%	15%
Solution a	3.75 ml	3.75 ml	7.5 ml	7.5 ml
Solution c	3.75 ml	5 ml	12.5 ml	15. ml
H_2O	7.4 ml	6.2 ml	9.9 ml	7.4 ml
TEMED	7 μl	7 μl	15 μl	15 μl
Solution e	56 μl	56 μl	112 μl	113 μl

A 10–12.5% acrylamide gel provides the optimal conditions for separation of the two polypeptide chains that constitute ATCase. To resolve a mixture of large and small polypeptides, one can use a gradient of acrylamide concentration. The casting stand must be filled with the polymerizing gel up to 4 cm from the cut-out part.

The 4.5% acrylamide upper gel has the following composition:

- Solution b: 2 ml
- Solution c: 1.2 ml
- H_2O: 4.7 ml
- TEMED: 5 μl
- Solution e: 75 μl

This mixture can be very cautiously deposited on top of the lower gel even before polymerization of the latter. By using a Pasteur pipette, one can proceed progressively without disturbing the surface of the lower gel. The comb is then placed between the two plates into the upper gel solution in order to form wells (Fig. 23). After about 2 hr of polymerization at room temperature (or about 30 min at 37°C), the gel is ready to use. It is sometimes necessary to prepare the gel 24 hr before use. This is the case, for instance, if one plans to transfer the proteins after electrophoresis, as will be described later. It appears that a delay between preparation of the gel and its use favors the removal of chemical species able to react with the proteins to be analyzed.

After polymerization, carefully remove the comb and the lower spacer, rinse the wells with water, and grease the gasket to ensure watertightness. Clamp the casting stand containing the gel onto the electrophoresis apparatus, the cut-out part being oriented toward the upper reservoir. Number the wells on the glass plate (Fig. 23).

Fill the two reservoirs with solution d, ensuring that the two ends of the gel are in contact with this buffer; remove the air bubbles from the lower part of the gel with the help of a syringe equipped with a bent needle. Add 5–6 drops of the indicator solution to the buffer present in the upper reservoir and mix until homogeneous. If the proteins are to be transferred after the electrophoresis, also add to the upper tank thioglycolic acid at a final concentration of 1 mM.

Samples

Each protein sample to be analyzed must be diluted with an equal volume of solution f. Heat the mixture for 1–2 min at 100°C as previously described. The maximum volume that can be deposited in a well is generally 30 μl. Hamilton syringes or automatic pipettes equipped with very narrow tips are convenient to gently layer the samples in the wells. Use the same standard protein solutions as those previously described.

Migration

Run the experiment with a current intensity of about 30 mA. Under these conditions, the migration takes about 3 hr.

Staining and Destaining

Cautiously remove the gel from the plates. To keep track of its orientation, cut its lower right corner. Stain for 30 min at room temperature and destain for about 2 hr, changing the solution several times during that period. These operations are performed in gently swirled covered flat boxes, which hold the methanol vapors, or in a fume hood. If the gel is to be dried for long-term storage, it must first be immersed for 6–12 hr in the solution for drying. It is then dried between cellophane sheets, under vacuum, using a specially devised apparatus.

Remarks

The Coomassie blue solution can be used about 10 times. To avoid the appearance of nonspecific stain spots on the gel during its first use, the solution should be allowed to settle several days before filtration through paper.

The limit of detection by Coomassie blue is about 0.5 μg of protein. Alternatively, one can use silver coloration (Merril *et al.*, 1981), which allows the detection of about 5 ng of protein.

Using slab gels, it is easy to transfer the proteins onto a membrane for which they have some affinity. Nitrocellulose membranes are widely used for this purpose. Polyvinylidene difluoride membranes (such as Immobilon-P from Millipore) have the important advantage of allowing the transferred proteins to be sequenced directly (Matsudaira, 1987). The transfer is made with an unstained gel, but the proteins can be subsequently stained on the membrane even if they are to be sequenced. It is sometimes possible to renature the transferred proteins after the removal of SDS. In such cases, certain functions of the proteins (ligand binding, enzymatic activity, etc.) can be tested directly on the membrane. In addition, after saturation of the membrane with bovine serum albumin or any other immunologically unrelated protein, they can be treated with specific antibodies directed toward the protein under study; this method is called *Western blotting* (Towbin *et al.*, 1979).

Analysis of ATCase and Its Subunits

Proceed as described above, using 1-mg/ml solutions of ATCase, catalytic subunits, and regulatory subunits.

Results

Analyze the results obtained as indicated in the section dealing with the electrophoresis in cylindrical gels.

Question 6. *Under what conditions can one use measurement of the intensity of the bands obtained after Coomassie blue staining to determine the stoichiometry of the polypeptide chains in ATCase? In other words, is it possible, by using this method, to reach the conclusion that the 34,000-dalton and 17,000-dalton chains are in the ratio 1:1 in the native enzyme?*

3. ANALYTICAL ULTRACENTRIFUGATION

Study of the hydrodynamic properties of macromolecules provides information about their shape, hydration, and molecular mass. Hydrodynamic methods include viscosimetry, osmometry, and analytical ultracentrifugation. Only the latter will be discussed here.

3.1. Principle of Analytical Ultracentrifugation

In analytical ultracentrifugation, one makes optical measurements on the sample during its sedimentation. For this purpose, the sample is placed in a cell equipped with windows, allowing measurement of the concentration of the macromolecule as a function of the distance from the axis of rotation. Knowing the geometry of the rotor, one can use optical references (internal and external) to accurately calculate distances within the cell, taking count of the enlargement generated by the optical system. The signal is recorded either on a photographic plate or a UV-recording scanner. This type of device is schematically presented in Fig. 24. A more complete description of the method can be found in several textbooks (Batelier, 1979; Cantor and Schimmel, 1980).

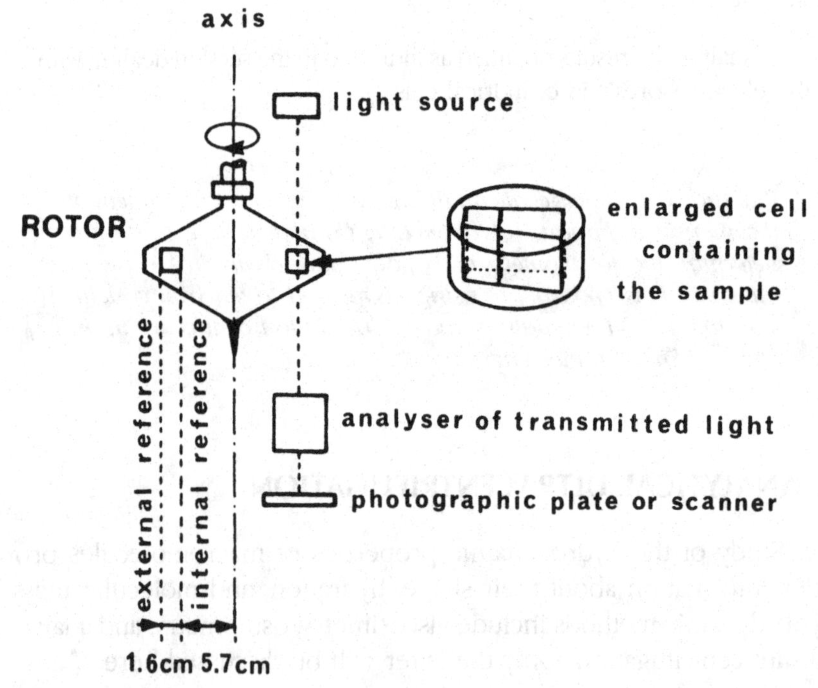

Figure 24. Principle of analytical ultracentrifugation. During centrifugation, the enlarged picture of the cell is recorded on the photographic plate or graph paper. The optical references (internal and external) allow the calculation of the enlargement. (Adapted from Lehninger, 1975, with permission.)

3.2. Sedimentation Velocity

3.2.1. Description of the Method

Figure 25A shows the variation of the macromolecule concentration profile during a sedimentation velocity experiment. At the beginning of the centrifugation, the macromolecule concentration in the sample is uniform. Then at times t_1 and t_2 of centrifugation, the macromolecule accumulates at the bottom of the cell, and its distribution defines three zones or regions as a function of the distance from the rotation axis (Fig. 25A)

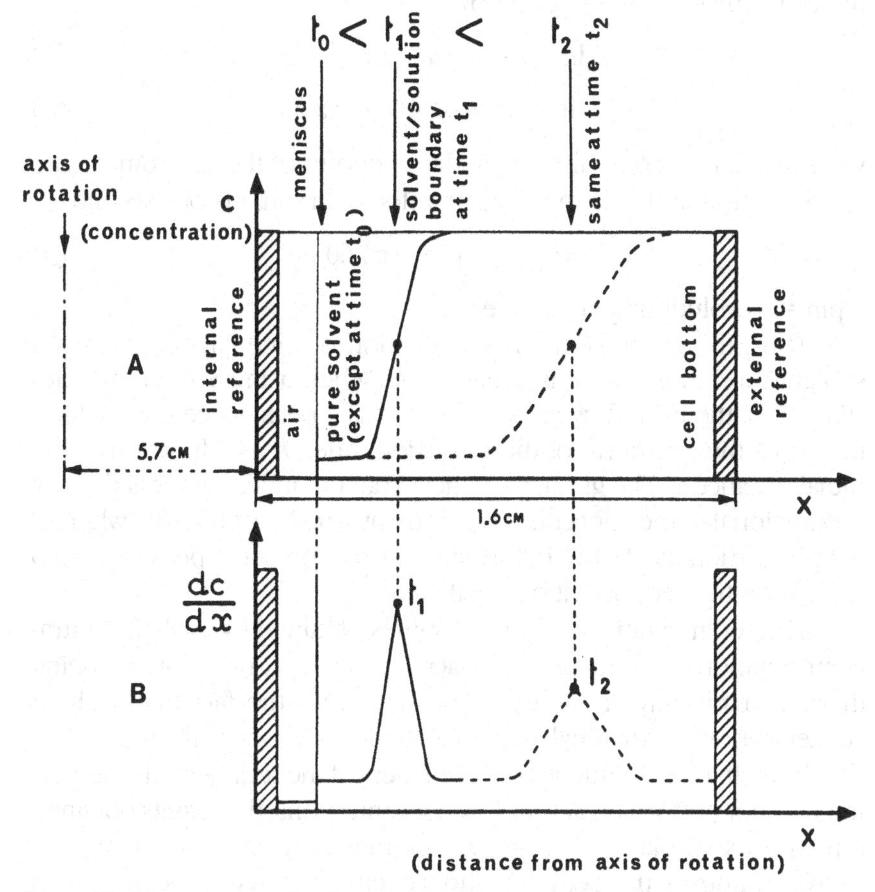

Figure 25. Measurement of sedimentation velocity. Variation of the macromolecule concentration along the cell is represented at two different times, t_1 and t_2: as it appears by a direct measurement of the optical density (A) and as it appears using Schlieren optics (B) (see text).

1. A plateau toward the bottom of the cell where the concentration of the macromolecule is more or less constant and close to the original concentration of the sample
2. A boundary region where the concentration of the macromolecule decreases progressively from the plateau value to zero
3. A supernatant region in which the macromolecule is no longer present (pure solvent)

The *sedimentation coefficient s* is measured by following the rate at which the inflection point of the concentration profile migrates during the centrifugation and is given by:

$$d(\ln r_M) = \omega^2 s dt \quad (7)$$

$$\ln r_M = \omega^2 st + \text{constant} \quad (8)$$

where r_M is the radial distance at the midpoint of the macromolecule concentration and ω is the angular velocity in radians per second.

$$\omega = (\text{rpm} \times 2\pi)/60 \quad (9)$$

(rpm = revolutions per minute)

Thus, the graph of $\ln r_M$ as a function of time (in seconds) is a straight line whose slope is equal to $\omega^2 s$. When using Schlieren optics (Fig. 25B), the boundary region appears as a peak, since this method measures the derivative of the refraction index n as a function of the radial distance x. The gradient of the refraction index dn/dx is directly proportional to the concentration gradient: $dn/dx = P(dc/dx)$, where P is a proportionality factor. In this method, the top of the peak is used to determine the sedimentation coefficient.

The sedimentation coefficient values obtained by analytical ultracentrifugation depend on the macromolecule concentration, being thermodynamically "nonideal." This is due to the fact that, at high concentrations, individual macromolecules interact with each other, thus lowering the sedimentation coefficient. If the sedimentation experiment is not performed at very low concentrations, the results obtained must be extrapolated to zero concentration in order to obtain the correct value of the sedimentation coefficient. When dealing with globular macromolecules, such as most of the water-soluble proteins, one considers that the sedimentation coefficient can be obtained directly only if the protein concentration is less than about 0.1 mg/ml.

At higher concentrations, it is necessary to extrapolate the value of s according to:

$$s = s^0(1 - Kc) \tag{10}$$

where s^0 is the sedimentation coefficient at infinite dilution, K is a constant that is a function of the size and shape of the macromolecule, and c is the concentration of the macromolecule. The sedimentation coefficient values obtained in different conditions of solvent or temperature may be compared only if they are expressed in terms of standard conditions. In the case of proteins, s^0 is converted into $s^0_{20,w}$, the value that would have been obtained in pure water at 20°C. This mathematical treatment of the results is not necessary in the following determinations, since the experiments reported were performed at 20°C, using a buffer whose density and viscosity are very close to those of water (i.e., $s^0 \cong s^0_{20,w}$).

3.2.2. *Sedimentation Velocity of ATCase and Its Isolated Subunits*

Native ATCase

The photographs shown in Fig. 26 will be used to analyze the sedimentation velocity of native ATCase. These photographs were obtained by centrifugation of an ATCase sample (7.5 mg/ml) at 60,000 rpm at 20°C in a 50-mM phosphate buffer (pH 7.2) containing 2-mM β-mercaptoethanol and 0.2-mM EDTA. To obtain the sedimentation coefficient, the photograph must be used as follows:

- Measure the distance (in centimeters) between the internal and external references (y in Fig. 26J)
- Measure the distance x between the *dn*/*dx* peak and the internal reference which is the closest one from the axis of rotation (x in Fig. 26J).
- This relative distance value is transformed into the real distance to the axis of rotation, knowing (as shown in Fig. 25) that the real distance between the two references in the cell is 1.6 cm, while the distance between the internal reference and the rotation axis is 5.7 cm. That is,

$$r_M = 5.7 + (1.6/y)\cdot(x) \tag{11}$$

- Present the results obtained in a form like that shown in Table 5
- On graph paper, represent the variation of $\ln r_M$ as a function of time. The slope $d \ln r_M/dt$ is equal to $\omega^2 s$ (equations 7 and 8)

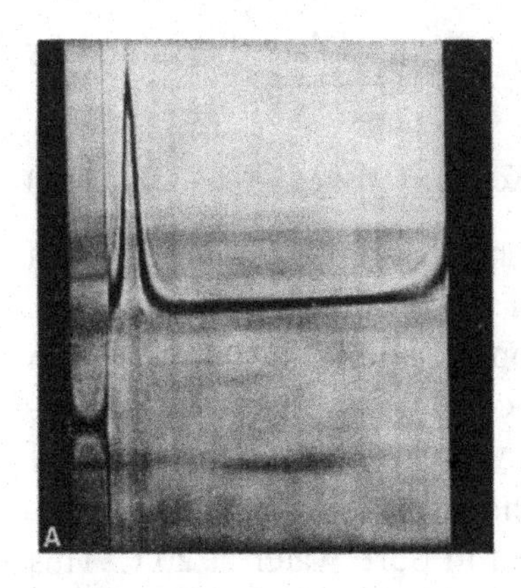
A

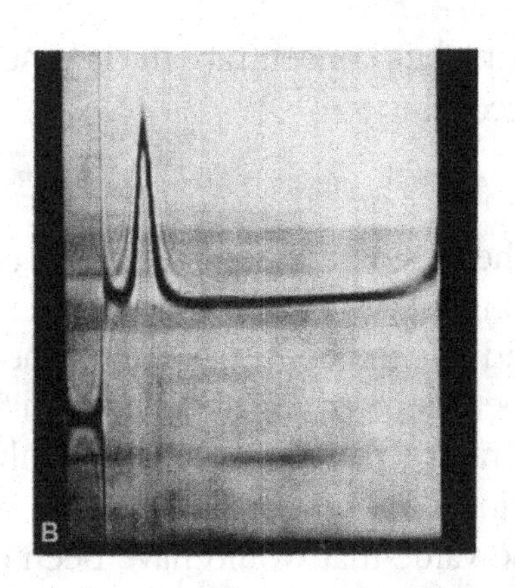
B

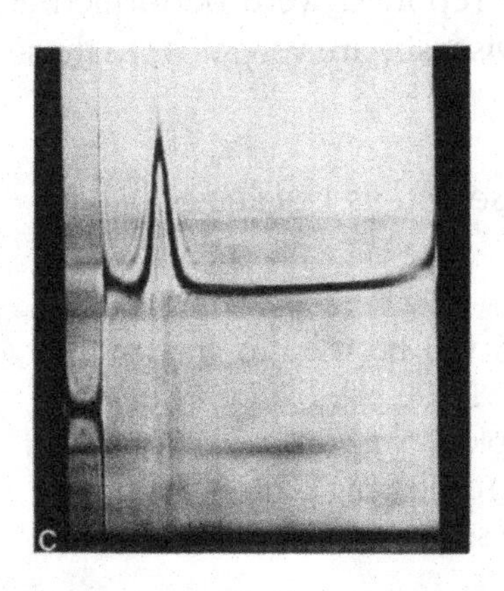
C

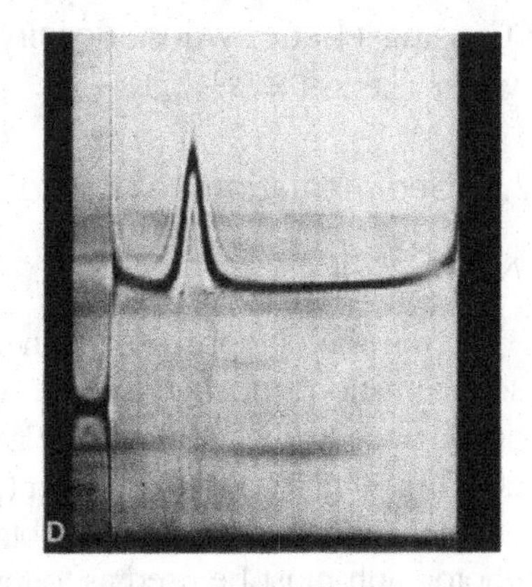
D

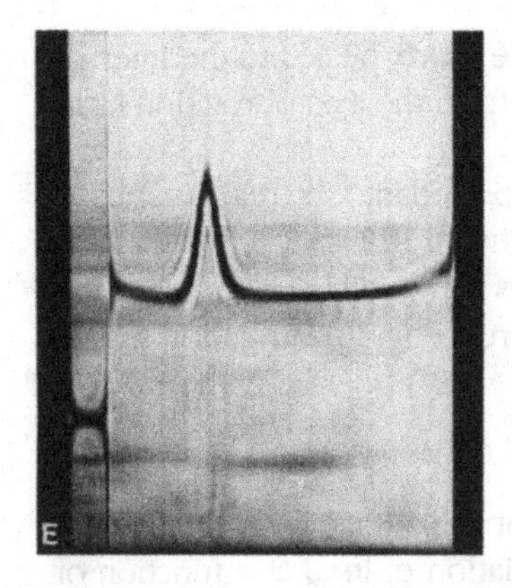
E

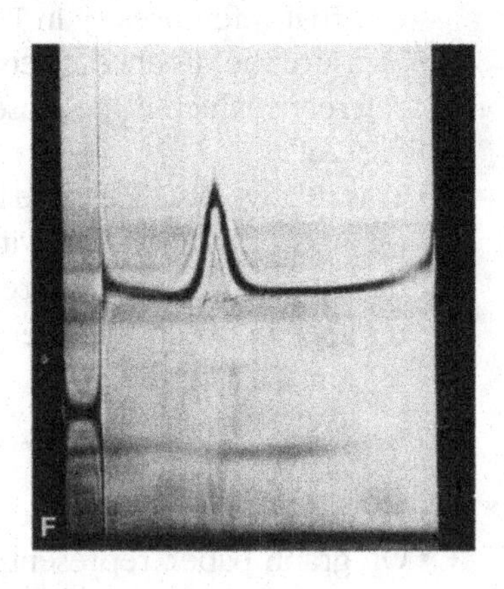
F

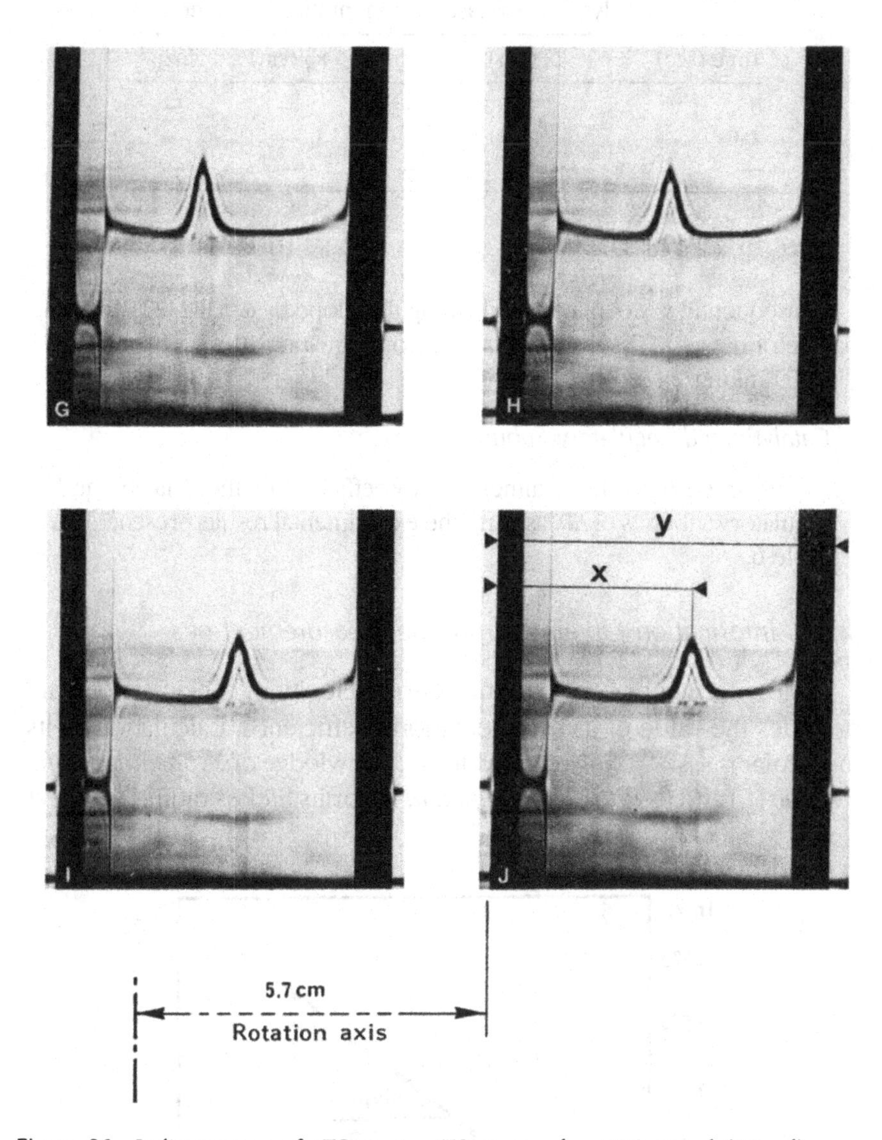

Figure 26. Sedimentation of ATCase. An ATCase sample at 7.5 mg/ml (0.3 ml) was centrifuged at 20°C at 60,000 rpm in 30-mM potassium phosphate buffer (pH 7.2) containing 2-mM β-mercaptoethanol and 0.2-mM EDTA, using a Beckman model E analytical ultracentrifuge. Photographs were taken every 4 min. The distance y between the internal and external references is given on J, as well as the distance x between the internal reference and the sedimentation peak. (A) t = 0; (B) t = 4 min; (C) t = 8 min; (D) t = 12 min; (E) t = 16 min; (F) t = 20 min; (G) t = 24 min; (H) t = 28 min; (I) t = 32 min; and (J) t = 36 min.

Table 5. Analysis of ATCase Sedimentation Photographs

Time (sec)	x (cm)	y (cm)	r_M (cm)	$\ln r_M$
0	—	—	—	—
240	—	—	—	—
—	—	—	—	—

Consequently, s is obtained by dividing this slope by ω^2 (Fig. 27). In the case of native ATCase, s^0 is obtained from equation 10 with K equal to 0.009 ml/mg.

Catalytic and Regulatory Subunits

To determine the sedimentation coefficient of the catalytic and regulatory subunits of ATCase, use the experimental results presented in Table 6.

3.2.3. *Information Obtained from the Measurement of s*

Measurement of the sedimentation velocity of a macromolecule provides the value of its sedimentation coefficient s. Calculation of its molecular mass M requires, in addition, knowledge of its *partial specific volume* ($\bar{v}$) and its *diffusion coefficient* D (or its Stokes radius; see next

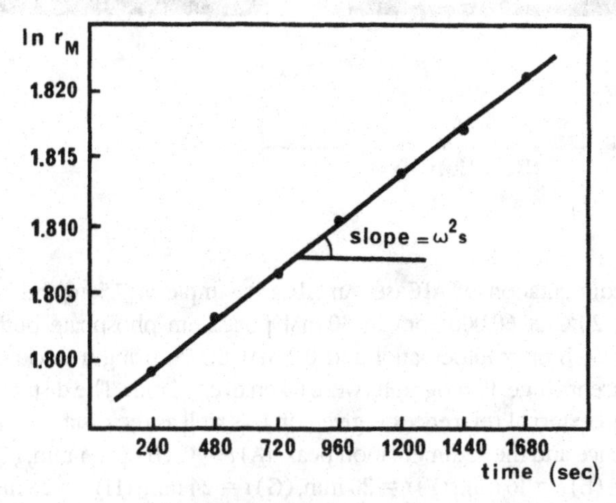

Figure 27. Determination of the sedimentation coefficient of a protein. In this experiment, in which the rate of rotation was 48,000 rpm, the value of s calculated from the slope of the curve is 6.02×10^{-13} sec.

Table 6. Sedimentation of the Catalytic and Regulatory Subunits of ATCase[a]

Time (min)	Cat. r_M (cm)	Reg. r_M (cm)
0	6.216	6.175
4	6.249	6.190
8	6.285	6.207
12	6.319	6.223
16	6.354	6.240
20	6.388	6.256
24	6.424	6.273

[a]For this analysis, 0.3-ml samples of 0.1-mg/ml solutions of catalytic (Cat.) and regulatory (Reg.) subunits were centrifuged at 20°C and 60,000 rpm in 50-mM potassium phosphate buffer (pH 7.2). Sedimentation was followed by UV absorption at 280 nm, a method that allows the use of dilute solutions. Consequently, in this case the factor *K* can be neglected.

section). The partial specific volume of a macromolecule is the increase of volume of a solution in which 1 g of this macromolecule is dissolved. It is measured in cubic centimeters per gram. The diffusion coefficient *D* is a measure of its Brownian diffusion in pure solvent and is expressed in square centimeters per second. That is,

$$s^0 = [MD^0(1 - \bar{v}\rho)]/RT \tag{12}$$

in which ρ is the density of the solvent, *R* is the universal gas constant, *T* is the absolute temperature, and D^0 the diffusion coefficient at infinite dilution. Since

$$D^0 = RT/Nf \tag{13}$$

where *N* is Avogadro's number and *f* is the frictional coefficient, one has:

$$s^0 = [M(1 - \bar{v}\rho)]/Nf \tag{14}$$

One can see that the sedimentation coefficient of a macromolecule depends not only on its molecular mass but also on its shape and hydration, which influence its frictional coefficient.

On the basis of very precise measurements of *s*, one can detect small conformational changes in proteins, induced for example by changes in pH or ionic strength or by ligand binding. This is the case for ATCase. In the presence of carbamylphosphate and succinate (a structural analog of the substrate aspartate), one observes a 3.6% decrease in the value of *s* for the native enzyme and a 1.1% increase in the case of the isolated catalytic subunits (Howlett and Schachman, 1977). These experiments, performed before crystallographic data were available, had already suggested that the binding of these ligands provokes a slight contraction of the catalytic subunits and swelling of the native ATCase molecule. These conformational changes in ATCase can also be studied by low-angle X-ray scattering in solution, a method which also provides information on the size and shape of macromolecules (Hervé *et al.*, 1985).

Concerning the determination of molecular mass (equation 12), the practical problems relate not to the precise measurement of s but rather to the determination of D, which is generally difficult. One prefers therefore either to measure the frictional coefficient (or, more exactly, the Stokes radius; see the discussion of column calibration in the next section) or to use another ultracentrifugation technique, sedimentation equilibrium, which is more accurate.

3.3. Sedimentation Equilibrium

The sedimentation equilibrium method allows direct measurement of the molecular mass of a macromolecule provided that its partial specific volume is known.

3.3.1. Principle

In the case of sedimentation equilibrium, one does not measure the rate at which a macromolecule moves in a high centrifugational field but rather measures the distribution of this macromolecule in a weak centrifugational field. In this case, the sample is submitted to two opposing effects: diffusion and sedimentation. When these two effects reach equilibrium (generally after 24–48 hr under the usual experimental conditions), the distribution of a given macromolecule as a function of the distance from the axis of rotation is directly proportional to its molecular mass. These experiments are performed using speeds of rotation lower than those used for determination of the sedimentation coefficient (for example, 10,000 instead of 60,000 rpm).

3.3.2. Measurement of Molecular Mass by the Interference Method

In the interference method, the light emitted by a pinpoint source is divided into two beams by two parallel slits located on the lower window of the centrifugation cell. These two beams are later recombined at the focus of a collimating lens, thus inducing the appearance of interference fringes. A change of refractive index at any level in the cell will lead to a deviation of the interference fringes. Since a change in macromolecule concentration causes a proportional change of the refractive index, the deviation of the fringes at any level is directly proportional to the macromolecule concentration at this level in the cell. Several methods can be used to determine the factor of proportionality. The base line of an interference fringe is given by its horizontal part on the right of the meniscus as it is presented in Figs. 28 and 29. This is a region of the cell where there is no longer any macromolecule but only solvent after sustained centrifugation.

3.3.3. Sedimentation Equilibrium of ATCase

The result of a sedimentation equilibrium experiment on ATCase is shown in Fig. 28. The photograph shows a series of parallel black-and-white fringes that are curved on the right side. To analyze the result, extend the linear part of one of these fringes toward the right side of the picture as indicated in Fig. 29. This is the base line. Then, draw the contour of the same fringe in its nonlinear part. One can see that in the case shown here, this contour deviates progressively from the base line starting at about $x = 9$ cm, x being the distance to the internal reference (Fig. 29). In the cell used in these experiments, this internal reference is located at 5.62 cm from the rotation axis. To determine the molecular mass of the protein, measure the deviation δ between the base line and the fringe as a function of the distance x. Collect the results as shown in Table 7, transforming the δ values into real concentrations and their logarithms. In the present case (Fig. 28), a fringe deviation of 10 mm corresponds to a protein concentration of 1.2 mg/ml. Finally, calculate r^2, knowing that 32.76 cm on the photograph corresponds to 1.68 cm real distance in the cell. This can be verified by using the bar representing 0.1 cm in Fig. 28. Thus

$$r = 5.62 + (1.68/32.76)x$$

Plot the variation of $\ln c$ as a function of r^2 as indicated in Fig. 30. In the case of a homogeneous solution (containing a single macromolecule species), one obtains a straight line whose slope is equal to:

$$d\ln c/dr^2 = [M(1 - \bar{v}\rho)\omega^2]/2RT \quad (15)$$

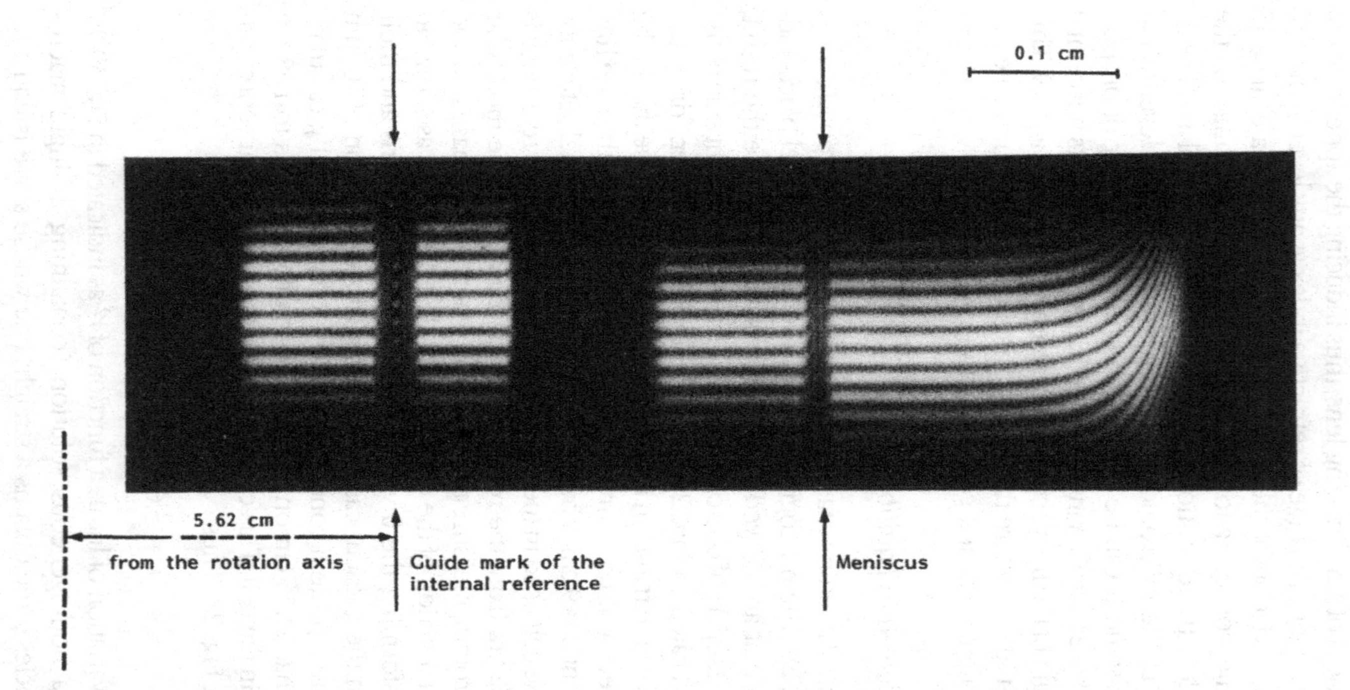

Figure 28. Result of sedimentation equilibrium experiments on native ATCase, using interference optics. A 100-μl sample of a 0.6-mg/ml solution of ATCase in 40-mM potassium phosphate buffer (pH 7) containing 2-mM β-mercaptoethanol and 0.2-mM EDTA was centrifuged 28 hr at 4°C and 13,000 rpm. The comparison of pictures taken at 4-hr intervals shows that equilibrium has been reached.

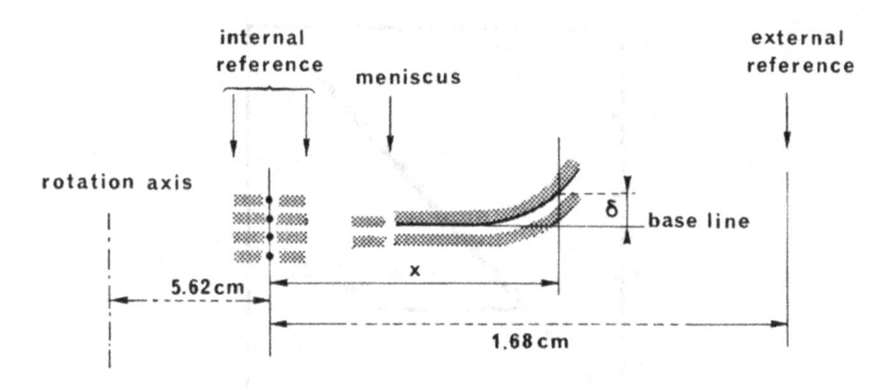

Figure 29. Method of analysis for the sedimentation equilibrium photographs.

where M is the molecular mass in grams per mole of the protein, $\bar{v}$ is its partial specific volume expressed in cubic centimeters per gram (in this case, $\bar{v} = 0.738$ cm^3/g), ρ is the density of the solvent (here 1 g/cm^3), R is the gas constant (83.144×10^6 erg °K^{-1} mol^{-1}), T is the temperature in degrees Kelvin (here 4°C = 277°K), and ω^2 is the square of the angular velocity, which is equal to:

$$[(\text{rpm} \times 2\pi)/60]^2$$

(here rpm = 13,000).

From the slope of the line obtained, calculate the molecular mass of ATCase.

Table 7. Analysis of a Native ATCase Sedimentation Equilibrium Photograph

x (cm)	δ (mm)	c (mg/ml)	ln c	r^2 (cm^2)
8.5	0.5	—	—	—
9	—	—	—	—
9.2	—	—	—	—
9.4	—	—	—	—
9.6	—	—	—	—
—	—	—	—	—

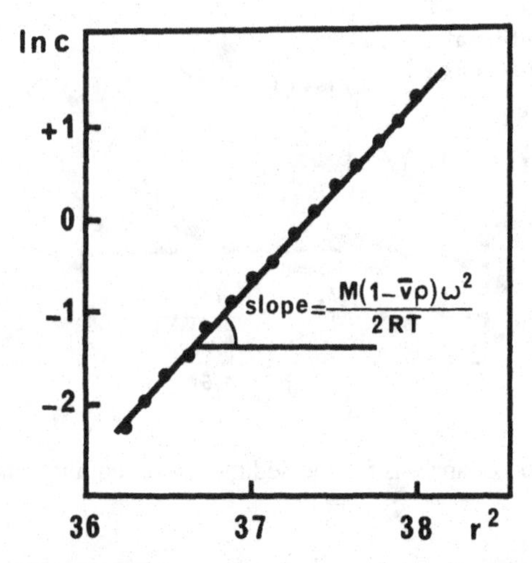

Figure 30. Determination of the molecular mass by sedimentation equilibrium.

4. SIZE EXCLUSION CHROMATOGRAPHY

The size of macromolecules can also be determined by size exclusion chromatography.

4.1. Principle

Chromatography through porous gels is used in biochemistry, mainly as a method of separating macromolecules of different sizes. This method is sometimes incorrectly called molecular sieving or gel filtration; contrary to what would occur with a sieve, in this method it is the larger molecules that are released first. The macromolecules whose dimensions are larger than that of the gel pores are completely excluded from these pores, whereas smaller molecules are able to enter, more or less deeply, into the pores as a function of their size and shape. Thus, the molecules are eluted from the gel in the order of decreasing size. Consequently, the expression ***size exclusion chromatography*** is more appropriate to designate this method.

Compared with other separation methods such as electrophoresis or ion exchange chromatography, size exclusion chromatography has the advantage of being mild; it does not perturb the structure of the macromolecules. Furthermore, it provides very reproducible results.

Contrary to what is often implied in the literature, proteins are not eluted only as a function of their molecular mass. Parameters such as their shape and hydration, and other factors such as direct interactions with the gel, can influence their elution. Thus, gel chromatography must be used with care for determining the physicochemical parameters of macromolecules. Attempts to directly determine the molecular mass of a protein by comparison of its elution with that of proteins of known molecular masses are not advised. This procedure can lead to errors larger than 50% (le Maire *et al.*, 1986). The true parameter than can be determined by this method is the ***Stokes radius***.

4.2. The Stokes Radius

On the basis of their hydrodynamic properties, proteins can be classified as being either globular, rod shaped, or random coiled.

The so-called globular proteins, although relatively spherical and compact, do not all have the same shape. This is easily seen by comparing the proteins whose three-dimensional structure has been determined by X-ray crystallography. Nevertheless, their behavior on gel chromatography can be compared in terms of the Stokes radius, a physicochemical parameter that takes into account their size, shape, and hydration. The Stokes radius of a hydrodynamic particle (such as a protein in solution) is defined as the radius of the sphere that would have the same frictional coefficient as this particle (Tanford, 1961)

$$f = 6\pi\eta R_S \tag{16}$$

where f is the frictional coefficient and η is the viscosity of the solvent expressed in poise. In the course of hydrodynamic measurements such as sedimentation velocity, viscosity, or gel chromatography, proteins carry a certain amount of solvent (water, in normal conditions). The less compact (random coil) or spherical (rod shaped) the proteins are, the more solvent they carry. When a protein rotates, the solvent stirring is larger if this protein is elongated than if it is spherical. The Stokes radius of a protein is thus the radius of the *equivalent hydrodynamic sphere* that would produce a stirring (frictional coefficient) identical to the one observed in the case of this protein in solution. The Stokes radius would be equal to the true radius only in the purely hypothetical case of a protein that is perfectly spherical and does not carry solvent molecules.

When the differences in shape and hydration of a series of proteins are not too large, use of the Stokes radius (R_S expressed in nanometers) allows all of these proteins to be compared as equivalent spheres. The calibration of a gel column with a series of proteins whose R_S values are known (Table 8) is then possible, and the combination of R_S and $s_{20,w}$ (measured as previously described) allows determination of the molecular mass of an unknown protein by using the following equation, which is derived from equations 14 and 16:

$$M = (6\pi\eta R_S N s_{20,w})/(1 - \bar{v}\rho) \tag{17}$$

Table 8. Useful Proteins for the Calibration of Columns for Size Exclusion Chromatography

Protein[a]	Origin	Molecular mass	$s_{20,w}$ (S)	R_S (nm)
Cytochrome *c*	Horse heart	12,400	1.83	1.7
Ribonuclease A	Beef pancreas	13,400	1.9	1.75
Myoglobin	Horse muscle	17,800	2.04	1.9
Trypsin inhibitor	Soja	22,100	1.3	2.2
Hemoglobin[b]	Human or bovine	32,000	3.03	2.4
β-Lactoglobulin	Cow milk	36,000	2.80	2.75
Ovalbumin	Chicken egg	43,000	3.55	2.8
Alkaline phosphatase	*E. coli*	86,000	5.9	3.3
Albumin	Bovine	66,000	4.15	3.5
Transferin	Human	85,000	4.9	3.6
Aldolase	Rabbit muscle	158,000	7.7	4.6
Catalase	Beef liver	240,000	11.1	5.2
Ferritin[c]	Horse spleen	≅800,000	≅55.	6.3
β-Galactosidase	*E. coli*	464,000	16.2	6.9
Thyroglobulin	Bovine	669,000	19.1	8.6

[a]These proteins are commercially available. However, some preparations contain several molecular species that are either impurities or dissociated subunits (for example, β-galactosidase). Molecular masses and sedimentation cofficients are taken from the literature (Siegel and Monty, 1966; Fasman, 1976; le Maire *et al.*, 1980, 1986). With the exception of ferritin, Stokes radii are deduced from $M(1 - \bar{v}\rho)$ obtained by sedimentation equilibrium and from the sedimentation coefficient, using equation 17. Values were obtained in buffers of ionic strength 0.05–0.2 and pH close to 7. These values can be used only if the columns are calibrated under similar conditions. Indeed, the presence of ionic detergent or urea in the buffer changes the R_S. Furthermore, some ionic strength and pH conditions favor the dissociation of multimeric enzymes (β-galactosidase, catalase).

[b]At the concentrations generally used in columns (≅0.1 mg/ml), hemoglobin is dissociated into dimers.

[c]The commercially available preparations of ferritin contain variable amounts of iron, and thus their molecular masses and sedimentation coefficients are not constant. Values given in the table refer to the Pharmacia preparation, which is virtually saturated with iron. In any case, since this cation is bound in the center of the macromolecule, its presence alters only slightly the value of the Stokes radius. Indeed, in the case of apoferritin (deprived of iron), the molecular mass of 475,000 and the $s_{20,w}$ of 17.4S give a Stokes radius of 6.1 nm, a value close to the one in the table. Actually, ferritin and apoferritin have very similar diameters when observed by electron microscopy.

This method is not very accurate but can be used with caution when the molecular mass cannot be measured directly by sedimentation equilibrium. This is the case, for instance, when only a small amount of the protein is available, which can be detected only on the basis of its enzymatic activity. This is also the case when this protein is not purified to homogeneity but can be monitored on the basis of its activity. In these cases, one can also obtain the ***relative sedimentation coefficient*** by centrifugation on sucrose gradients. However, some possible causes of error must be taken into account (Tanford *et al.*, 1974).

The Stokes radius can be used to obtain information other than the molecular mass. Comparison of R_S with the minimum radius of a protein (R_{min}) can provide information about its degree of asymmetry and of hydration. Although these two parameters are contained in the concept of the Stokes radius, their relative impacts cannot be distinguished without the help of additional experiments. By definition, R_{min} is the radius of a protein assumed to be perfectly spherical and dry, that is:

$$V = \frac{4}{3}\pi R^3_{min} \tag{18}$$

where V is the volume occupied by 1 mole of dry protein. Since $V = M\bar{v}$, then

$$R_{min} = \sqrt[3]{\frac{3M\bar{v}}{4\pi N}} \tag{19}$$

Because of the asymmetry and hydration of the protein, R_S is always larger than R_{min}. The experimental data show that R_S/R_{min} is equal to about 1.2 for soluble globular proteins but can be much higher in the case of very asymmetrical proteins such as myosin ($R_S/R_{min} = 2.3$) or highly hydrated proteins such as albumin dissolved in the presence of 6 M guanidine–HCl or 8-M urea ($R_S/R_{min} > 3$).

Example: Ovalbumin ($M_r = 43{,}000$) has a partial specific volume of 0.748 cm^3/g and a Stokes radius of 2.8 nm at pH 7 (Table 8). The volume of 1 mole of this protein is given by $V = M\bar{v}$, that is, 32,164 cm^3. On this basis, equation 19 gives $R_{min} = 2.34$ nm. Consequently, $R_S/R_{min} = 1.19$, a value indicating that ovalbumin is a globular protein. In the presence of SDS and β-mercaptoethanol, used during analysis by gel electrophoresis under denaturing conditions, one obtains for ovalbumin a Stokes radius of 6.31 nm. Taking into account the amount of SDS bound to the protein under these conditions (1.4 g/g; see Section 2) and the SDS partial specific volume (0.87 cm^3/g), one can calculate V = 32,164 cm^3 + (1.4 × 43,000 × 0.87) = 84,534 cm^3. From this value one obtains R_{min} = 3.22 nm and an R_S/R_{min} ratio of 1.9, which indicates that under these conditions ovalbumin is no longer a globular protein. Additional experiments lead to the conclusion that, as mentioned earlier (Section 2), the deviation from 1.2 is due to a greater asymmetry of the SDS–ovalbumin complex rather than to a higher degree of hydration. On the contrary, in the presence of 6 M guanidine–HCl, ovalbumin has an R_S of 6.28 nm. To a first approximation, in this case one can use the R_{min} of ovalbumin under nondenaturing conditions, since only a very low amount of guanidine is bound to the protein. Thus, R_{min} = 2.34 and R_S/R_{min} = 2.69. Again, it appears that under these conditions ovalbumin is no longer globular. In this case, however, this phenomenon is due not to a higher asymmetry, but to a loss of compactness associated with a higher hydration of the protein. It is possible to distinguish between these two possibilities (rigid rod or random coil) by determination of the radius of gyration of the protein (Tanford, 1961) or comparison with a series of proteins whose hydrodynamic properties are known.

Considering the definition of the Stokes radius, one might expect the calibration of a gel to be universal, that is, valid whatever the class of protein to be analyzed: elongated proteins such as myosin, fibrinogen, SDS- or urea-denatured proteins, glycoproteins, membrane proteins solubilized in detergents. Unfortunately, this is generally not the case (le Maire *et al.*, 1989a), which means that additional caution in the use of analytical gel chromatography is warranted.

4.3. Column Calibration

Accurate calibration of the chromatography columns to be used is very important. This calibration is based on the relationship that exists between the ***distribution coefficient*** and the Stokes radius.

4.3.1. Measurement of the Distribution Coefficient

The elution volume of a protein (V_e) is generally linked to two intrinsic parameters of the gel and of the column, that is, the *void volume* (V_o) and the total volume (V_t), according to the equation:

$$K_d = (V_e - V_o)/(V_t - V_o) \qquad (20)$$

where K_d is the distribution coefficient of this protein. V_o is the volume of liquid outside the gel particles (Fig. 31). It is experimentally measured by the elution of very large molecules unable to penetrate the pores of the gel. V_t is the volume accessible to the small molecules that can penetrate all of the pores. V_e is the volume accessible to intermediate size molecules able to partially enter the pores. Thus,

$$V_o < V_e < V_t$$

The distribution coefficient K_d of a protein is the ratio of the volume that is accessible to this protein inside the gel pores to the volume accessible to very small molecules. In principle, for a given type of gel (Sepharose 6B, for example), the K_d of a protein is constant whatever the amount of gel used, that is, even if V_o, V_e, and V_t vary.

Some authors use the term K_{av}, whose definition is slightly different from that of K_d:

$$K_{av} = (V_e - V_o)/(V_{col} - V_o) \qquad (21)$$

where V_{col} is the volume of the column, given by the surface of the cross section of the column multiplied by the gel height. The relationship with K_d is

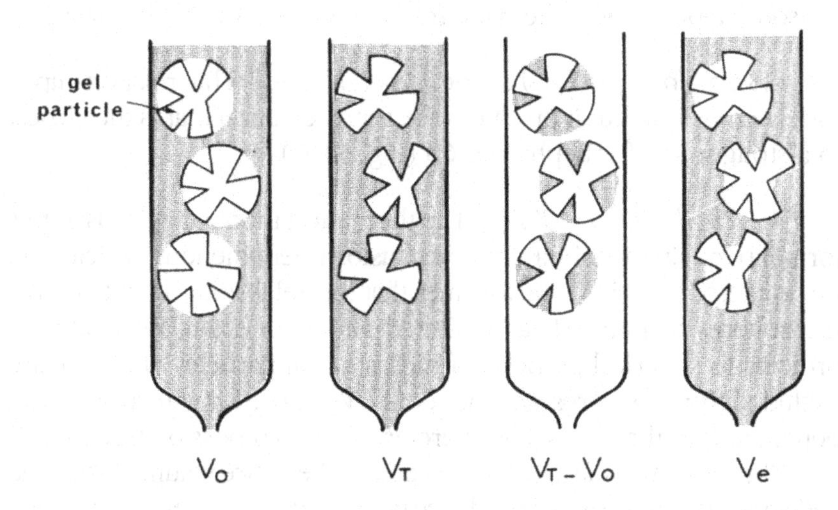

Figure 31. Schematic representation of the different volumes involved in the process of size exclusion chromatography. It is assumed that the gel contains pores of different sizes. Shaded areas correspond to the volumes indicated. (Adapted from le Maire, 1987.)

$$K_d = (V_e - V_o)/[(V_{col} - V_m) - V_o] = (V_e - V_o)/(V_t - V_o) \quad (22)$$

where V_m is the volume of the gel matrix. Thus, the difference between K_{av} and K_d is the volume of the gel matrix, which generally is very small. For example, Sepharose 6B contains about 6% agarose and 94% water.

$V_t - V_o$ is sometimes called V_i or V_S; this is the volume of the solvent contained inside the gel pores (Fig. 31).

4.3.2. Calibration Curves

The graphic representation of the relationship between the Stokes radius and K_d (or K_{av}) is called the *calibration curve*. It can be obtained in several ways. The simplest and most accurate involves the direct representation of the variation of R_S as a function of K_d and does not imply any physical model. Alternatively, plotting R_S as a function of $1 - K_d$ provides a positively oriented curve. This representation gives a sigmoidal curve (Fig. 32,■). Other types of representations tend to linearize the curves, either through a purely mathematical operation or on the basis of a particular theoretical model of size exclusion chromatography. Such linear representations facilitate the calibration. Since a straight line is defined by only two points, two standard proteins should theoretically be sufficient to calibrate a column. In fact, experimentation shows that there is presently no unique linear graph that would be valid for all types of gels (le Maire *et al.*, 1987, 1989a,b). However, some of these graphs provide linear relationships for some types of gel. The most frequently used are the following.

4.3.2a. $\log R_S = f(K_d)$ or $\log R_S = f(1 - K_d)$. This relationship is purely empirical, but it produces the representation that is the closest to a straight line for Sepharose 6B (Fig. 32, ○).

4.3.2b. $R_S = f(1 - K_d^{1/3})$. The theoretical model that leads to this formula explains how, by a simple exclusion phenomenon, particles can be separated (Fig. 33). One assumes that the gel is equivalent to empty half-spheres of radius L; as indicated above, one can assimilate the proteins to spherical particles of radius R_S. All particles of $R_S > L$ are excluded from the pores, and they define V_o. Those for which $R_S < L$ can penetrate into the pores, and there are then two possibilities:

The very small particles have access to the entire volume inside the half-spheres. Their path or residence time inside the pores is maximum, and their elution volume is V_t. The internal volume of the pores is $V_t - V_o = \frac{1}{2}(\frac{4}{3}\pi L^3)$.

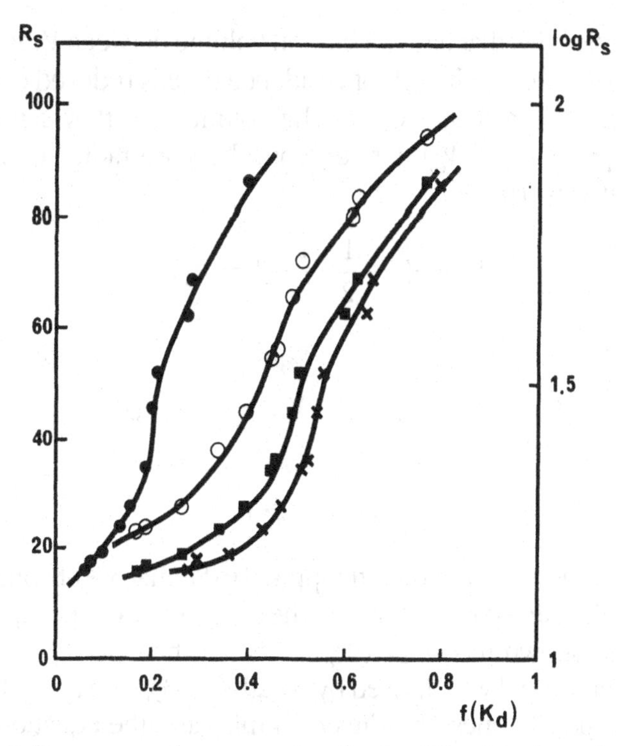

Figure 32. Different types of representations for the calibration of a Sepharose 6B gel column. $f(K_d)$ on the abscissa corresponds to $1 - K_d$ (○ and ■), $1 - K_d^{1/3}$ (●), $(-\log K_d)^{1/2}$ (X). The ordinate $\log R_S$ corresponds to the curve ○—○.

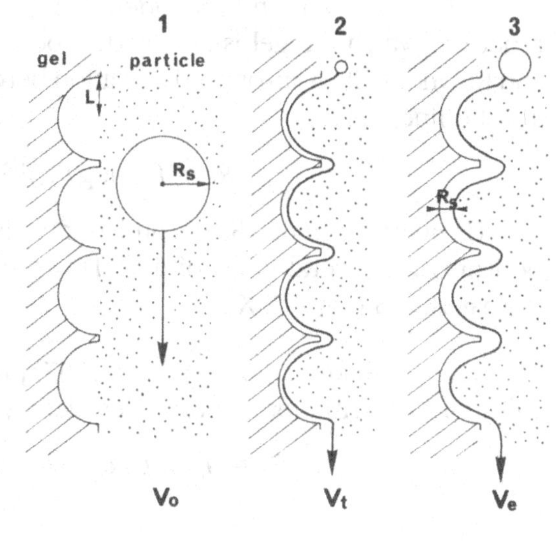

Figure 33. Simplified model for size exclusion chromatography. One assumes a single pore size (half-spheres of radius L). Particle 1 is excluded, and its elution volume is V_o. Particle 2 has access to the entire volume of the pores, and its elution volume is V_t. Particle 3, whose center of gravity cannot approach the pore wall closer than a distance equal to its radius, has access only to a limited region of the pores. Its elution volume is V_e, and the exclusion volume (undotted area) has a thickness of R_S, (i.e., its Stokes radius).

The molecules that have an elution volume between V_o and V_t also enter the pores, but their path or residence time is reduced compared with that of the small molecules. The volume that they can explore inside the pores ($V_e - V_o$) derives from the pore radius minus their Stokes radius R_S. That is,

$$V_e - V_o = \frac{1}{2} \cdot \frac{4}{3}\pi(L - R_S)^3 \tag{23}$$

and

$$K_d = (V_e - V_o)/(V_t - V_o) = (L - R_S)^3/L^3 \tag{24}$$

or

$$R_S = L(1 - K_d^{1/3}) \tag{25}$$

If this model could be directly applied to Sepharose 6B, one should obtain a straight line passing through the origin with a slope equal to the pore size; as shown in Fig. 32 (●), this is not the case.

A similar model was devised by Porath (1963), who assimilated the pores to identical cones of radius L; in this case, the equation is

$$R_S = \frac{L}{2}[1 - (K_d/k)]^{1/3} \tag{26}$$

where k is a constant. For some types of gels, equations 25 and 26 give straight lines (Siegel and Monty, 1966).

4.3.2c. Laurent and Killander (1964). Using a different physical model in which the gel is assumed to be a network of rods, Laurent and Killander (1964) proposed a relation between K_{av} and R_S which can also be linearized:

$$R_S = f(-\log K_{av})^{1/2} \tag{27}$$

Using this formula, Siegel and Monty (1966) obtained linear relationships for several types of gels. However, this is not the case for Sepharose 6B (Fig. 32, X).

4.3.2d. Ackers (1967). Ackers (1967) proposed a model based on a Gaussian distribution of pore sizes, deriving the following equation:

$$R_S = a_o + b_o\, erf^{-1}(1 - K_d) \tag{28}$$

where a_o and b_o are the pore dimensions corresponding to the maximum of the distribution and the standard deviation of that distribution, respectively; erf^{-1} is the inverse function of the probability integral. This relation has been widely used, but it was shown recently that its derivation was not mathematically correct (le Maire *et al.*, 1987). Furthermore, the distribution of the pore sizes was not Gaussian in all of the cases investigated, and equation 28 does not give a straight line for Sepharose 6B (le Maire *et al.*, 1980, 1987).

4.3.2e. *Fractal Model.* The weakness of the models described above probably arises in part from the assumptions made regarding the structure of the gels and the distribution of pore sizes. If one assumes that the porous media used in size exclusion chromatography are surface fractals, one obtains the following equation (le Maire *et al.*, 1989b; Brochard *et al.*, 1989);

$$\ln R_S = \ln L + [\ln(1 - K_d)]/(3 - D_f) \tag{29}$$

where D_f is the fractal dimension of the gel and L is the maximum radius of the pores. For some standard gels which are in fact of heterogeneous structure, this relationship cannot be applied. In the case of high-performance liquid chromatography (HPLC) gels (see below), a straight line is obtained when $\ln R_S$ is plotted as a function of $\ln(1 - K_d)$, and these gels can indeed be characterized by a fractal dimension D_f and a maximum radius of the pores L (le Maire *et al.*, 1989b).

4.3.3. *Experiments*

Size exclusion chromatography can be used for several purposes:

1. To separate proteins of different sizes, for example, ATCase and its subunits.
2. As an analytical method for determining the Stokes radius of proteins, a parameter which, combined with the sedimentation coefficient obtained by methods previously described, allows the approximate molecular mass to be calculated.
3. The ratio R_S/R_{min} of a protein provides information about its shape and its hydration.

4.3.3a. *The Reservoir.* The reservoir (Fig. 34) contains 300 ml of the buffer that will be used to eluate the column. In this case the buffer is

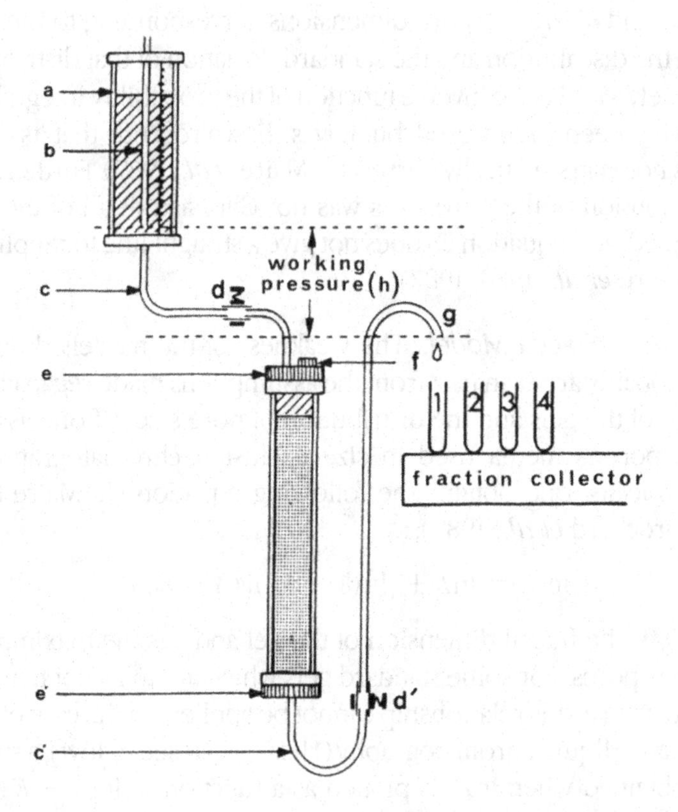

Figure 34. Column for gel chromatography. a, Reservoir of eluant; b, air tubing of the Boyle–Mariotte flask; c and c′, inlet and outlet Teflon tubings (the outlet tubing comprises a safety loop that prevents accidental drying of the gel); d and d′, clamps placed on Tygon tubings; e and e′, adaptors; f, air-draining cork; h, working hydrostatic pressure.

10-mM *N*-tris(hydroxymethyl)methylaminomethanesulfonic acid (TES) (pH 7.5)–100-mM KCl–1-mM sodium azide. This solution does not absorb at 280 nm. The ionic strength is high enough to minimize protein–protein and gel–protein electrostatic interactions. Azide is a bactericide. Verify that there are no air bubbles in the Teflon tubing (c in Fig. 34) and that the eluant can flow freely. Clamp this tube (d) at the point where Teflon is joined to the soft plastic, Tygon (d). Empty the air tube (b) with a syringe fitted to Tygon tubing. The reservoir is then equivalent to a Boyle–Mariotte flask. The working pressure (h) applied to the column, which is measured as the distance between the lower part of the air tubing and the column outlet, will determine the flow rate of the column. Thus, the working pressure will remain constant whatever the level of eluant in the reservoir.

4.3.3b. The Column. Use a Pharmacia-type column 1.6 cm in internal diameter and 100 cm in height. The column must be vertical. Attach the packing funnel to the top, pour 10 ml of buffer, and verify that the buffer flows freely, with no air bubbles in the outlet tubing c′. Clamp this tube, leaving the buffer at a height of about 1 cm in the column. Fix the end of the outlet tubing about 30 cm below the top of the packing funnel, thus establishing the working pressure for filling the column.

The same pressure will be used later for elution of the column. The choice of this pressure is based on the nature of the gel and the dimensions of the column. When size exclusion chromatography is used as an analytical method, the flow rate should not be too high; for the type of gel and column described here, the flow rate should be 5–10 ml/hr.

4.3.3c. Types of Gel Used. The gel used is chosen on the basis of the sizes of the macromolecules to be separated. For a given type of gel, the separation efficiency of a column can be estimated as N_p, the number of *theoretical plates* per column:

$$N_p = 5.54 \times \left(\frac{\text{elution volume (in grams)}}{\text{mid-height width of the peak (in grams)}}\right)^2 \quad (30)$$

This value is generally determined by elution of a small molecule such as β-mercaptoethanol. The greater N_p, the more efficient the column. This value is frequently expressed for a 1-m-long column (N_p/column length in meters). For Sepharose 6B and TSK 3000 SW, which is used in HPLC, these values are, respectively, 2000 and 15,000 theoretical plates/m. The separating power of the gels is also sometimes expressed in terms of height equivalent of theoretical plates (HETP), the inverse of N_p/m:

$$\text{HETP} = m/N_p \quad (31)$$

The gel used in the experiments described here is Sepharose 6B (Pharmacia). It has rather large pores that allow the entry of molecules with large Stokes radii. The number 6 indicates the approximate percentage of agarose in the gel beads, the remainder being water.

Thus, Sepharose 4B and 2B have larger pore sizes (Table 9). Sephadex gels (cross-linked dextran) have smaller pore sizes. Sephacryls are copolymers of dextran–acrylamide and allow fast flow rates with a good resolution.

All of these gels are called standard gels, (in comparison with size exclusion HPLC gels (le Maire *et al.*, 1987). The HPLC gels are generally sold in ready-for-use columns. They require the use of high pressure and a more complex apparatus that includes high-precision pumps and spectrophotometric detectors. Though more costly, HPLC has several advantages: it is 10–40 times faster, and its resolution is about ten times higher in terms of number of theoretical plates. For HPLC gels, the calibration procedure is the same as that described below (le Maire *et al.*, 1986) except that V_o, V_t, and V_e can be measured directly from the elution profile monitored by the spectrophotometric detector.

4.3.3d. Gel Preparation. Unlike Sephadex, the commercially available Sepharose is provided ready to use, already hydrated, and in suspension in water. However, before the columns are prepared, it is recommended that the gel suspension be diluted with 0.5–1 volume of elution buffer to decrease the viscosity and so avoid trapping air bubbles in the gel. For the same purpose, the diluted gel is deaerated under vacuum for about 1 hr at the temperature at which the column will be run. In the case described here (1.6 × 100-cm column) 250 ml of diluted gel will occupy about 80% of the volume of the column, the rest being buffer (Fig. 34).

4.3.3e. Packing of the Column. The homogenized gel suspension must be gently poured onto the side of the packing funnel, whose volume must be at least 50 ml in order to allow packing the 250-ml gel

Table 9. Features of Various Sepharose Gels

Gel	Approximate wet bead diam (μm)	Useful fractionation range for proteins ($M_r \times 10^{-6}$)	Agarose (%)
Sepharose 6B	45–165	0.01–4	6
Sepharose 4B	45–165	0.06–20	4
Sepharose 2B	60–200	0.07–40	2

suspension in a single step. Proceed slowly in order to avoid air bubbles, especially at the bottom of the column. When all of the gel has been poured into the column, the level of buffer must be adjusted to the top of the packing funnel. It is important that the top of the gel never be dry. To establish the working pressure of 30 cm, immediately remove clamp d′ (Fig. 34). The gel is kept in the column by a net located in the bottom adaptor, while the eluant flows dropwise. It is important to ensure that the gel packs homogeneously in the column; therefore, the level of buffer in the packing funnel must be maintained constant. After a few hours, the gel will be packed. Its upper level must be about 10 cm from the top of the column. The surface of the gel must be flat and horizontal. If this is not the case, the upper part of the gel must be stirred and then allowed to settle again. Remove the packing funnel and connect the column to the buffer reservoir, whose height must be adjusted in order to maintain the working pressure (Fig. 34). Remove clamp d.

It is then necessary to equilibrate a newly packed column with 2–3 volumes of buffer. Before setting up the fraction collector, verify that the flow rate of the column is 5–10 ml/hr.

4.3.3f. Weighing of the Tubes for Fractionation of the Eluate. With the type of gel and column described here, a satisfactory resolution is obtained only if the volume of the fractions collected is smaller than about 1.8 ml. A further decrease in the volume of the fractions does not significantly improve the result. Since the total volume of the column is about 150 ml, 100 tubes are sufficient to collect all molecules loaded on the column, provided that the volume of the fractions is not less than 1.5 ml. With a flow rate of about 7 ml/hr, set the fraction collector to about 15 min per tube.

To ensure high accuracy, the position of the elution peak of a protein (V_e) is expressed in *mass*, not in volume. It is therefore necessary to calculate the mass of buffer already eluted when the protein peak appears. In practice, the tubes are weighed before and after column elution and V_e is obtained by subtraction. The flow rate of the column is relatively constant (Table 10) provided that there is no variation of the working pressure, temperature, or appearance of air bubbles in the tubings. Therefore, it is sufficient to weigh the tubes by series of ten; then the mass corresponding to an elution peak is obtained by interpolation as indicated in Table 10 and Fig. 35.

Table 10. Determination of Eluant Mass during Gel Chromatography[a]

Rack number	Tube number	Mass of tubes with their racks (g)	Mass of filled tubes with their racks (g)	Differences (g)	Cumulative mass (g)
10	1–10	155.41	171.01	15.6	15.6
11	11–20	154.83	170.32	15.49	31.06
12	21–30	155.35	170.77	15.42	46.51
13	31–40	151.69	167.19	15.50	62.01
14	41–50	154.87	170.38	15.51	77.52
15	51–60	155.09	170.63	15.54	83.06
16	61–70	153.98	169.52	15.54	108.60
17	71–80	154.88	170.47	15.69	124.19
18	81–90	153.97	169.61	15.64	139.83

[a]Values correspond to the experiment presented in Fig. 35.

4.3.3g. Column Calibration

Determination of Void Volume (V_o) and Total Volume (V_t)

The first part of the experiment allows the determination of V_o and V_t by using blue dextran and dithiothreitol (DTT), respectively. Blue dextran is a mixture of large polymer with an average molecular mass of several million. Most of this blue dextran sample is thus excluded from the gel and is eluted as a sharp peak. The number of grams corresponding to this peak is V_o (about tube 25 under the conditions described here). As a result of the size heterogeneity of the blue dextran sample, a second widespread peak appears later, which is not taken into account. DTT (M_r = 154) can enter into all the pores of the gel, and its elution peak gives V_t (about tube 80 under the experimental conditions described here).

Procedure

The calibration mixture is prepared from a 50-mg/ml blue dextran solution 100-mg/ml DTT solution in distilled water. A 100-μl sample of each is added to 800 μl of elution buffer containing, in addition, either 5 mg of sucrose or 150 μl of glycerol, in order to raise the density. To obtain good resolution, it is important not to add an excess of glycerol; the viscosity of the sample should not be more than twice that of the eluant. After stirring, this mixture must be clear. In general, one does not add the protein sample to this mixture, since it might react with DTT or aggregate with the blue dextran.

When gel chromatography is used as an analytical method, loading

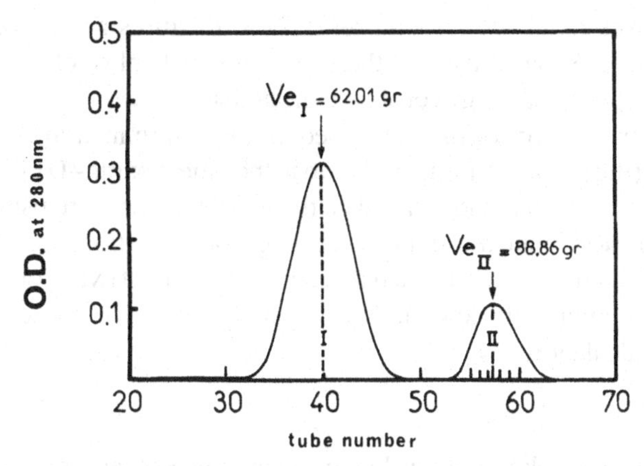

Figure 35. Elution profile of a mixture of two proteins. The peak of elution of protein I is located in tube 40. According to Table 10, its cumulative mass (V_{e_I}) is 62.01 g. The peak of protein II elutes between tubes 57 and 58, more precisely in the theoretical tube 57.3. According to Table 10, $V_{e_{II}}$ is thus 77.52 + (7.3 × 1.554) = 88.86 g.

of the sample is a crucial step. The sample must be gently deposited on the surface of the gel without perturbing it. This is often done by dipping the end of a syringe or micropipette containing the sample through the eluant that covers the surface of the gel. Since agarose gels are soft, one should be careful not to inject the sample inside the gel. It is convenient to cover the surface of the gel with a small filter paper disk whose diameter is that of the column. This surface must be smooth and horizontal. When the sample has been loaded, connect c to the column, reopen d and d′ (Fig. 34), and start the fraction collector.

When about 100 fractions have been collected (this takes about 24 hr) clamp d and d′. Weigh the tubes and calculate the cumulative masses as shown in Table 10. Measure the optical density of each tube at 280 nm, using a spectrophotometer and 1.5-ml quartz cuvettes.

Results

Draw the elution profile on chart paper and calculate V_o and V_t as indicated in Fig. 35.

Determination of the V_e of a Series of Standard Proteins

It is now necessary to calibrate the column with about a dozen proteins whose Stokes radii are known (Table 8). These determinations are performed by using two proteins at a time, chosen in such a way that their elution peaks do not overlap (implying a difference of 1–2 nm in their Stokes radii; see Appendixes). These mixtures are prepared by dissolving 3–5 mg of each of these proteins in 1 ml of elution buffer containing sucrose or glycerol as indicated above.

Continue as previously described for the determination of V_o and V_t but using the protein samples rather than the blue dextran–DTT mixture. From the results obtained, calculate the K_d of each protein, using equation 20. To accelerate this important process of calibration, one can use a series of identical columns, each receiving a mixture of two proteins. This procedure implies the knowledge of V_o and V_t for each of these columns for the calculation of K_d.

Results

Using the Stokes radius values given in Table 8 plot R_S as a function of K_d according to the different graphs, shown in Fig. 32. These graphs will be used to determine the Stokes radii of ATCase and its subunits.

4.3.3h. Determination of the K_d of ATCase and Its Subunits. Samples of ATCase (0.2 mg) or catalytic subunits (0.1 mg) together with 5 mg of the two marker proteins (see Appendixes) are chromatographed as indicated above. After elution, the absorption is measured at 280 nm; ATCase and the catalytic subunits are localized on the basis of their enzymatic activity. For this purpose, run one of the assays described in Chapter 5, using 100-μl samples in the case of the radioactive aspartate assay or 5-μl samples in the case of the colorimetric assay, adjusting the final volumes as indicated in Tables 12 and 13 (see Chapter 5).

The regulatory subunits are devoid of catalytic activity. They are detected on the basis of their absorption at 230 nm in an elution experiment in which a 0.2-mg sample is used in the ***absence*** of any other proteins.

Determine the V_e of these different proteins as indicated in Fig. 35 and calculate the corresponding K_d by using equation 20.

Results

Estimate the efficiency of the column (N_p, equation 30, N_p/m).

Question 7. *From the calibration curves $R_S = f(K_d)$, calculate the Stokes radii of ATCase and its catalytic and regulatory subunits.*

From these results and the values of $s_{20,w}$ previously obtained (Section 3.2), use equation 17 to calculate the molecular masses, knowing that the solvent viscosity (η) is 0.0102 poise, its density (ρ) is 1 g/cm^3, and $N = 6.02 \times 10^{23}$. Assume that the partial specific volume ($\bar{v}$) of C and R is

0.738 cm^3/g, as is that of ATCase. Remember that Stokes radii normally expressed in nanometers must be transformed into centimeters in equation 17.

In the case of ATCase, compare the molecular mass value obtained with the more accurate result provided by sedimentation equilibrium (Section 3.3).

Discuss the sources of error in the determination of R_S.

Question 8. *Calculate the R_S/R_{min} ratios and discuss the shape of ATCase and its subunits.*

Question 9. *Using the results from gel electrophoresis in the presence of SDS, a technique that provides the molecular mass of the individual catalytic and regulatory chains, calculate the chain composition of the catalytic and regulatory subunits.*

Question 10. *On the basis of the results from previously described experiments (which ones?), we know that ATCase (E) contains in mass more C than R. Is it possible to establish the formula $E = 2(C_3)3(R_2)$? Are there other possible combinations?*

Table 11 describes the different physicochemical techniques and their suitability to the study of protein molecular parameters.

Table 11. Comparison of Different Techniques for Determining the Physicochemical Parameters of Proteins

Technique	Parameters measured[a]	Description and limits of the technique[b]	Test of homogeneity[c]	Requirement for: Absolute measurement of concn[d]	High concn[e]	Large amount of protein[f]	Accessibility of the apparatus[g]	Time[h]
Sequencing	M_r	See below[i]	No	No	No	No	Easy	Variable
Sedimentation equilibrium	M_r (R_S if s known)	Chap. 4, 3	Good	No	No	No	Poor	≃24 hr
Sedimentation velocity	s (M_r if R_S known)	Chap. 4, 3	Good	No	No	No	Poor	≃1 hr
Diffusion of translation	R_S (M_r if s known)	Chap. 4, 3	Rather good	No	No	No	Poor	≃1 hr
Viscosity	$[\eta]$ ($R_{S\eta}$ if M_r known)	See below[j]	No	Yes	Yes	Yes	Poor	≃1 hr
Size exclusion chromatography	R_S, apparent (M_r if s known)	Chap. 4, 4	Good	No	No	No	Very easy	15 min–24 hr
Zonal centrifugation	s, apparent (M_r if R_S known)	See below[k]	Limited	No	No	No	Easy	≃10 hr
Light scattering (static)	M_r, apparent	See below[l]	Poor	Yes	Yes	Yes	Limited	≃1 min
Light scattering (dynamic)	R_S (M_r if s known)	See below[l]	Good	No	Yes	No	Poor	≃4 hr
Low-angle X-ray scattering	M_r, R_G, v_b, shape	See below[m]	Low	Yes	Yes	No	Limited	5 min–5 hr
Neutron scattering	M_r, R_G, v_b, shape	See below[m]	Low	Yes	Yes	No	Poor	10 min–5 hr
SDS–gel electrophoresis	M_r, apparent	Chap. 4, 2	Very good	No	No	No	Very easy	≃3 hr
Osmotic pressure	M_r	See below[n]	No	Yes	Yes	Yes	Very easy	3–4 days

[a] The ***apparent*** measurements imply a standardization with a series of reference proteins. The use of such a calibration alters the accuracy of the results, since some atypical proteins are not directly comparable to the standard proteins used. We distinguish here between accuracy and precision. The ***accuracy*** refers to the deviation from the real value, whereas ***precision*** refers to the reproductibility of the measurement and to the standard deviation of the experimental value and thus determines the number of significant figures of an experimental value. Techniques that allow an ***absolute*** measurement (sedimentation equilibrium, for instance) are always to be preferred, although they are less frequently used in practice than are size exclusion chromatography, zonal centrifugation, and gel electrophoresis. It is assumed in this table that the partial specific volumes are known. These values are easily calculated from the amino acid composition of the protein studied (Cohn and Edsall, 1943). More complex cases such as detergent-solubilized membrane proteins were analyzed by Steele *et al.* (1978) and Møller *et al.* (1986). Symbols: $[\eta]$, intrinsic viscosity; $R_{S\eta}$, viscosity-based Stokes radius; R_G, radius of gyration; v_b, hydrated volume.

[b] Complete descriptions of these techniques and their limitations are given in textbooks such as those of Tanford (1961), Eisenberg (1976), and Cantor and Schimmel (1980) or in various research articles referred to in the relevant chapter or below. It should be noted that the combined use of several techniques to study the same protein provides more accurate and extensive information. This is the case, for instance, for the combined use of sedimentation equilibrium and low-angle X-ray scattering (le Maire *et al.*, 1981; Tardieu *et al.*, 1981).

[c] Some techniques can provide indications as to whether a sample is homogeneous or not (e.g., containing monomers and oligomers). In the case of sedimentation equilibrium, for instance, the homogeneity of the starting solution can be estimated first from the plot of $\ln c = f(r^2)$, which gives a straight line for a homogeneous solution (Chapter 4, Section 3), or from the yield of recovery (le Maire *et al.*, 1978); finally, experiments at several speeds will provide the same molecular mass if the sample is homogeneous. In sedimentation velocity experiments, one can analyze the shape of the boundary (Møller *et al.*, 1986). This test of homogeneity is always very important.

[d] Absolute measurement of the protein concentration is necessary for only a limited number of techniques. A relative measurement is enough in the other techniques, making them more accessible (Chapter 4, Section 1).

[e] Some techniques require a "high" concentration of protein. This is a disadvantage, since at high concentrations some proteins tend to aggregate as a result of protein–protein interaction. It is arbitrarily considered in this table that a "high" concentration means above 1 mg/ml.

[f] The amount of protein required is here taken as high if above 1 mg. In some cases, only a very limited amount of pure protein is available.

[g] The concept of accessibility is considered here in a broad sense that includes the availability of the equipment and the complexity of its use. This notion can change rapidly. For instance, sedimentation equilibrium becomes increasingly difficult if one plans to use a model E Beckman or Centriscan MSE, but the availability of an airfuge or TL-100 Beckman has changed this situation (Pollet *et al.*, 1979; Minton, 1989). Similarly, osmotic pressure measurements are increasingly used thanks to the development of simple procedures (Vérétout *et al.*, 1989).

[h] Very approximate values depending on the type of apparatus used (conventional or Synchrotron X-ray sources or other parameters such as protein concentration, molecular mass, volume, temperature, and time for equilibration). One must also consider that in some cases it is necessary to make measurements at several protein concentrations in order to extrapolate the results to zero concentration (Chapter 4, Section 3).

[i] DNA sequencing (e.g., Berger and Kimmel, 1987) provides the molecular mass of the polypeptide chain of a protein with absolute accuracy but ignores the possible posttranslational modifications such as glycosylation, which in some cases increases significantly and variably the molecular mass of the functional form of the protein. For instance, the glycosylation of transferin increases its mass by about 6%. Binding of hemes or cofactors is not taken into account by this method, nor is the state of oligomerization of the protein. Only the hydrodynamic methods can answer these questions.

From the amino acid sequence of a protein, one can make predictions about its secondary structure. At present, the accuracy of these predictions does not exceed 65% of correct structure. Secondary, tertiary, and quaternary structure predictions are discussed by Jaenicke (1987).

[j] For globular proteins, the viscosity-based Stokes radius R_{S_η} coincides with the friction-based Stokes radius R_S obtained by diffusion methods. For random coils or elongated proteins, this is no longer true: $R_{S_\eta} > R_S$ (Horiike *et al.*, 1983; le Maire *et al.*, 1989a).

[k] This technique, like size exclusion chromatography, does not allow the ***direct*** measurement of molecular masses, since, the shape and hydration of proteins influence their movements (see discussion of sedimentation velocity in Chapter 4, Section 3). Errors as large as 100% may be found if molecular mass is directly assessed. However, one can obtain relative sedimentation coefficients. In that case, one must assume that the studied protein behaves as do the standard proteins used in the presence of sucrose; that is, it does not bind sucrose but binds about 0.3 g of water per gram of protein (le Maire *et al.*, 1981; Tardieu *et al.*, 1981). One assumes also that all of the partial specific volumes are the same. Such an assumption is even more dangerous when one is studying detergent-solubilized membrane proteins (Tanford *et al.*, 1974). A more accurate but more complex method allows one to take into account differences in $\bar{v}$ (McEwen, 1967).

[l] Light scattering (static) permits observation of the dimerization or aggregation of a protein as a function of time. It can provide R_G only for very large macromolecules such as viruses. Quasi-elastic light scattering (Dubin *et al.*, 1970) is a dynamic method based on the analysis of the Brownian motion of molecules in solution. It offers several practical advantages for the measurement of diffusion coefficients. However, as with all methods that involve light scattering, it is highly susceptible to the presence of large-size impurities or aggregates. The protein concentration required depends on the size of the protein; for example, one needs about 1 mg/ml for a molecular mass of 100,000 and about 0.5 mg/ml for a molecular mass of 300,000.

[m] With respect to v_b and molecular shape, the macromolecule under study must be homogeneous in terms of density; otherwise, a more complex investigation at variable contrast is necessary. Low-angle X-ray scattering (Luzzati and Tardieu, 1980; Tardieu *et al.*, 1981) is more sensitive to the shape of the macromolecule than neutron scattering (Zaccai and Jacrot, 1983), which is more sensitive to the internal fluctuations of density. Determination of molecular mass relies highly on the accuracy of $\bar{v}$ in X-ray scattering and much less in neutron scattering.

[n] At very high protein concentrations this technique also provides the interaction parameters which give information about charge, excluded volume, etc. (Parsegian *et al.*, 1986; Vérétout *et al.*, 1989). Interaction parameters can also be obtained by other methods, such as low-angle X-ray scattering.

Chapter 5

Enzymatic Catalysis and Regulation

1. THEORETICAL ASPECTS

The rate of an enzymatic reaction is measured as the amount of product, P, formed as a function of time or, alternatively, as the decrease of the amount of substrate, S. The rate of reaction is usually determined under experimental conditions in which the amount of S or P is directly proportional to time. These conditions are met only at the beginning of the reaction, when the concentration of P is small compared with the concentration of S, that is, the initial velocity.

The rate of reaction, V, varies as a function of a series of parameters: (1) substrate concentration, $[S]$; (2) enzyme concentration, $[E]$; (3) temperature, T; (4) pH; (5) pressure, P; (6) ionic strength, μ; and (7) concentrations of inhibitors, $[I]$, or activators, $[A]$.

Thus, V is a complex function, and it is necessary to study its variation as a function of a single parameter while maintaining all others constant. In each case, by following the reaction as a function of time, one can determine the conditions under which initial velocities can be measured.

1.1. Michaelian Enzymes

1.1.1. Variation of the Reaction Rate as a Function of Substrate Concentration

A simple enzymatic reaction in which the substrate, S, is transformed into the product, P, can be described by the following equation:

$$E + \underset{k_2}{\overset{k_1}{\rightleftarrows}} ES \underset{k_4}{\overset{k_3}{\rightleftarrows}} E + P \tag{32}$$

where E is free enzyme, ES is the enzyme–substrate complex in which the substrate is bound to the active site of the enzyme, and k_1–k_4 are rate constants of the different partial reactions.

Equation 32 accounts for the fact that any reaction, enzymatic or not, is theoretically reversible and characterized by an equilibrium constant. It applies to more complex situations involving several substrates or products, such as

$$S \rightleftarrows P_1 + P_2$$
$$S_1 + S_2 \rightleftarrows P$$
$$S_1 + S_2 \rightleftarrows P_1 + P_2$$

The theory of enzyme kinetics developed by Brown and Henri (Henri, 1903) and Michaelis and Menten (1913) can be applied satisfactorily in most cases. According to this theory, when $[S]$, the concentration of substrate, is much larger than that of the enzyme, $[E]$, the concentration of ES is constant, the rates of formation and disappearance of this complex being equal. From the law of mass action

$$k_1[E][S] + k_4[E][P] = k_2[ES] + k_3[ES]$$

Under the conditions of initial velocity, $[S]$ is virtually constant and $[P]$ is negligible, which means that

$$k_4[E][P] = 0$$

Consequently:

$$k_1[E][S] = (k_2 + k_3)[ES]$$

$$\frac{[E][S]}{[ES]} = \frac{k_2 + k_3}{k_1} = K_M$$

where K_M is the *Michaelis constant*.

In the presence of a large excess of substrate, the enzyme is entirely in the form of ES and the initial velocity of the reaction is maximal (V_m). Thus,

$$V_m = k_3[E_t] \tag{33}$$

where E_t is the total concentration of enzyme. For all the other concentrations of substrate, the rate of reaction is given by

$$V = k_3[ES] \tag{34}$$

since

$$[E] = [E_t] - [ES] \tag{35}$$

the combination of these equations gives

$$\frac{[S]}{k_3[ES]}(V_m - V) = K_M$$

or

$$V = \frac{V_m[S]}{K_M + [S]} \tag{36}$$

the usual form of the *Michaelis–Menten equation*. In this equation, if $V = V_m/2$

$$\frac{1}{2} = \frac{[S]}{K_M + [S]} \quad \text{and} \quad K_M = [S]$$

Thus, the Michaelis constant is equal to the concentration of substrate at which the rate of reaction is half the maximal velocity. It is expressed in concentration units. K_M and V_m are essential characteristics of an enzyme.

It is important to realize that K_M expresses the balance between formation and disappearance of the *ES* complex, but is not simply the dissociation constant of *S* from *ES*. Only when k_3 is much smaller than k_2 can K_M be equated to this dissociation constant. In that particular case

$$K_M = \frac{k_2}{k_1} = \frac{[E][S]}{[ES]}$$

This situation is actually encountered in many cases where K_M is then a measure of the affinity of the enzyme for its substrate. K_M values are usually found to be in the range of 10^{-6} M (high affinity) to 10^{-2} M (low affinity).

In the case of cooperative enzymes like ATCase, the plot of V as a function of $[S]$ is more complex (see below).

1.1.1a. Experimental Determination of V_m and K_M. V_m and K_M are determined by measuring the variation of initial rate, V, of reaction as a function of substrate concentration, $[S]$.

1.1.1b. Graphical Representations. Michaelis Plot. A Michaelis plot is the direct representation of V as a function of $[S]$ (Fig. 36). The curve obtained is a hyperbola whose asymptote is V_m. For this reason, V_m and K_M (measured for $V = V_m/2$) cannot be precisely determined, and mathematical methods were devised to linearize the rate equation (equation 36), as described in the following sections.

Double-Reciprocal Plot of Lineweaver and Burk (1934). This representation is based on the inverse of equation 36

$$\frac{1}{V} = \left(\frac{K_M}{V_m}\right)\left(\frac{1}{[S]}\right) + \frac{1}{V_m} \tag{37}$$

When $1/V$ is plotted against $1/[S]$, a straight line is obtained whose slope is equal to K_M/V_m and which intersects the ordinate for $1/V_m$ and the abscissa for $-1/K_M$ (Fig. 37).

This representation suffers from a statistical bias because the rates measured at low concentrations of substrate have an overweighted influence on the slope of the curve. Consequently, it is necessary to make an adequate choice of substrate concentration and to use complementary representations.

Eadie–Hofstee Plot (Eadie, 1952; Hofstee, 1952). Equation 36 can be written:

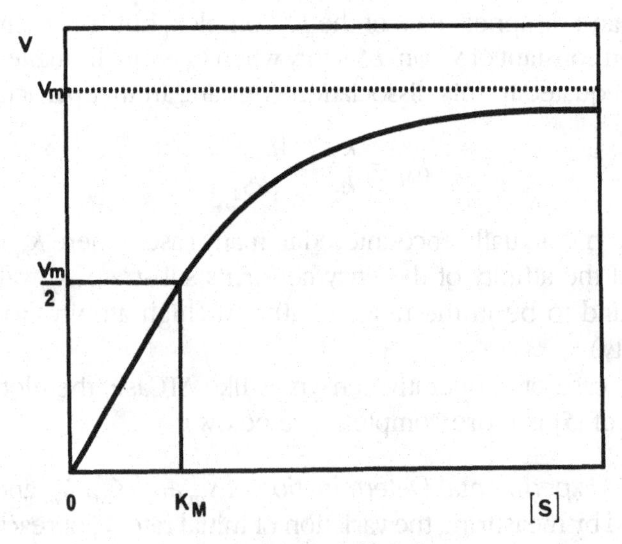

Figure 36. Michaelis plot.

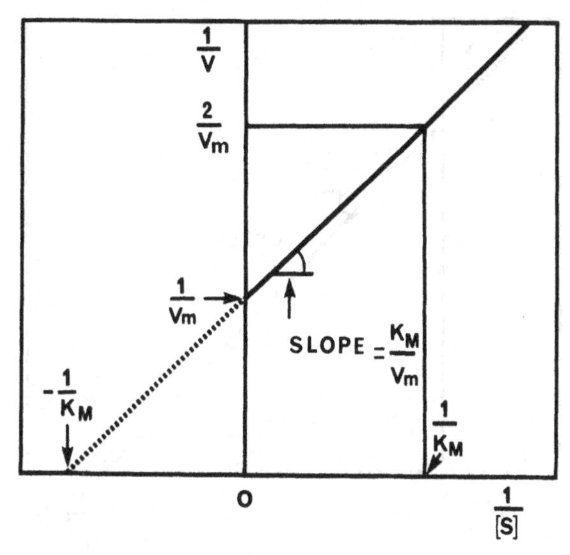

Figure 37. Double-reciprocal plot of Lineweaver and Burk (1934).

$$V(K_M + [S]) = V_m[S]$$

which, after dividing by $[S]$, gives:

$$V = V_m - \frac{V}{[S]}K_M \tag{38}$$

Thus, if one plots V as a function of $V/[S]$ (Fig. 38), a straight line is obtained whose slope is equal to $-K_M$ and which intersects the ordinate at $V = V_m$.

This is the best plot for determining non-Michaelian behavior.

Hanes Plot (Hanes, 1932). Equation 36 can also be written

$$\frac{[S]}{V} = \frac{K_M}{V_m} + \frac{1}{V_m}[S]$$

On plotting $[S]/V$ as a function of $[S]$ (Fig. 39), a straight line is obtained whose slope is equal to $1/V_m$ and which intersects the ordinate for K_M/V_m. This is the easiest way to obtain a linear representation of the experimental data with Michaelian enzymes.

Plot of Eisenthal and Cornish-Bowden (1974). The graphical determination proposed by Eisenthal and Cornish-Bowden (1974) eliminates the statistical errors associated with the previously described mathe-

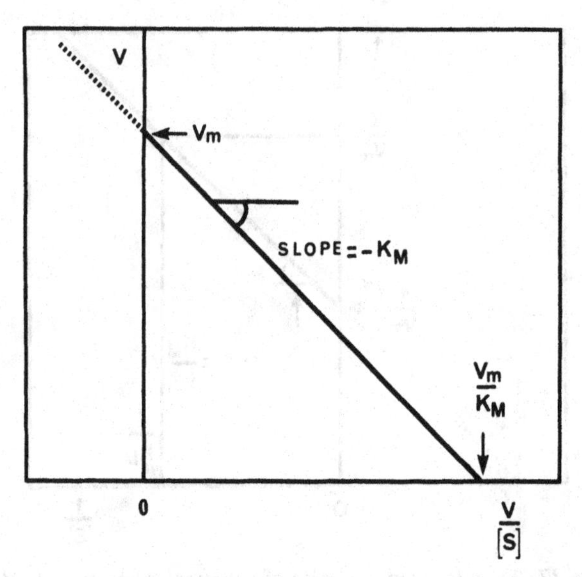

Figure 38. Eadie–Hofstee plot (Eadie, 1952; Hofstee, 1952).

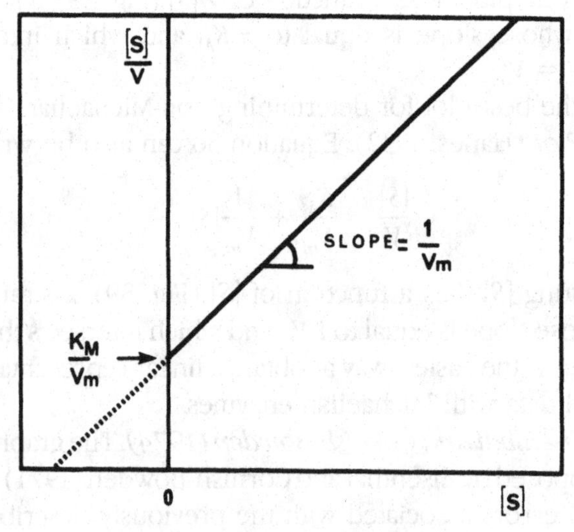

Figure 39. Hanes plot (Hanes, 1932).

matical treatments of the Michaelis–Menten equation. It involves plotting directly each value of V on an ordinate and the corresponding value of $[S]$ on the negative part of the abscissa and then drawing a straight line between these values (Fig. 40). The series of lines obtained intersects at a point whose coordinates are V_m and K_M (Fig. 40A). As a consequence of the experimental errors, in practice this is not the case.

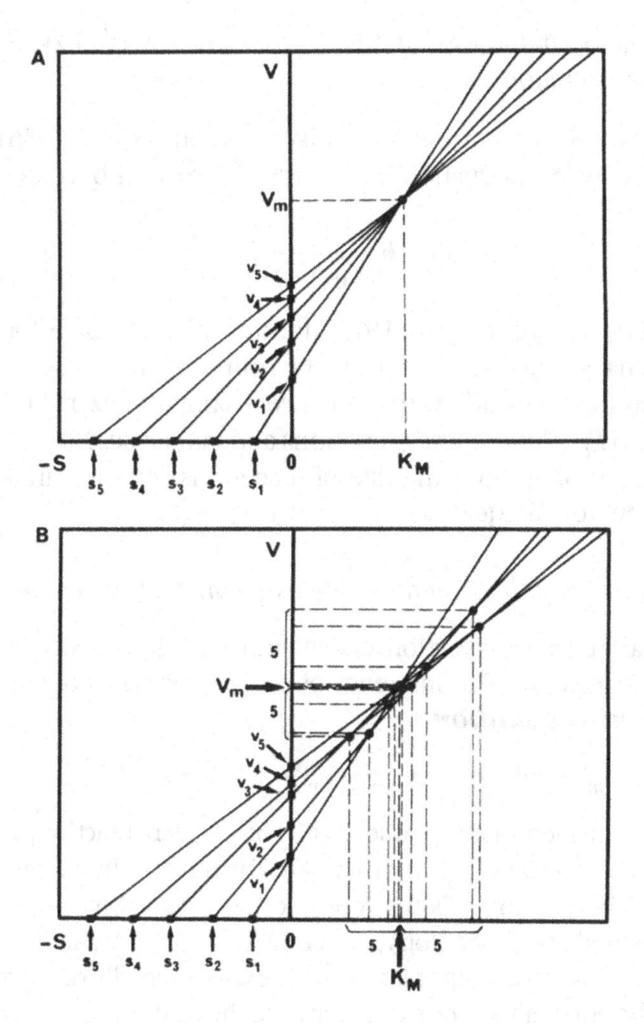

Figure 40. Determination of V_m and K_M by the graphical method of Eisenthal and Cornish-Bowden (1974). The filled squares on the abscissa and the ordinates are the concentrations of S used and the corresponding reaction rates, respectively. (A) Set of theoretical curves; (B) experimental determination of K_M and V_M from experimental data (see text). (Used with permission of A. Cornish-Bowden.)

Therefore, one analyzes the straight line by pairs. Each intersection is considered to provide an estimate of V_m and an estimate of K_M. The median (i.e., the middle) value from each series is the best estimate of V_m and K_M. If there is an even number of values, the median is taken as the mean of the middle two estimates (Fig. 40B).

This is the best way of determining V_m and K_M once Michaelian behavior has been established by the Eadie plot.

1.1.2. Variation of the Reaction Rate as a Function of Enzyme Concentration

If V_m is replaced in the Michaelis equation (equation 36) by its expression given in equation 33, the rate of reaction becomes

$$V = k_3 \frac{[S]}{K_M + [S]}[E_t] \tag{40}$$

As long as $[S]$ is large compared to K_M for this substrate, or as long as $[S]$ does not change significantly during the course of the reaction, the rate of reaction varies linearly with the concentration of enzyme (V is proportional to E_t). Under these conditions and in the presence of a given concentration of enzyme, the rate of reaction is constant, an absolute requirement for the quantitative estimation of $[E_t]$.

1.1.3. Variation of the Reaction Rate as a Function of Temperature

Like any chemical reaction, enzymatic catalysis is sensitive to variations of temperature. This influence of temperature is accounted for by the ***Arrhenius equation***

$$V_m = Ae^{-E/RT} \quad \text{or} \quad \log V_m = \log A - (E/2.3RT) \tag{41}$$

where A is the constant characteristic of a given reaction, R is the universal gas constant (8.314 j/mole/°K), T is the absolute temperature in degrees Kelvin, and E is the free energy of activation in joules.

In general, the plot of $\log V_m$ as a function of $1/T$ (Arrhenius plot) is linear in the range of temperature that is experimentally explored. The slope of the curve allows one to determine the free energy of activation of the enzymatic reaction (Fig. 41). Obviously, deviations from linearity are observed at high temperatures at which the enzyme is inactivated. The consequence of this phenomenon is that each enzyme exhibits an

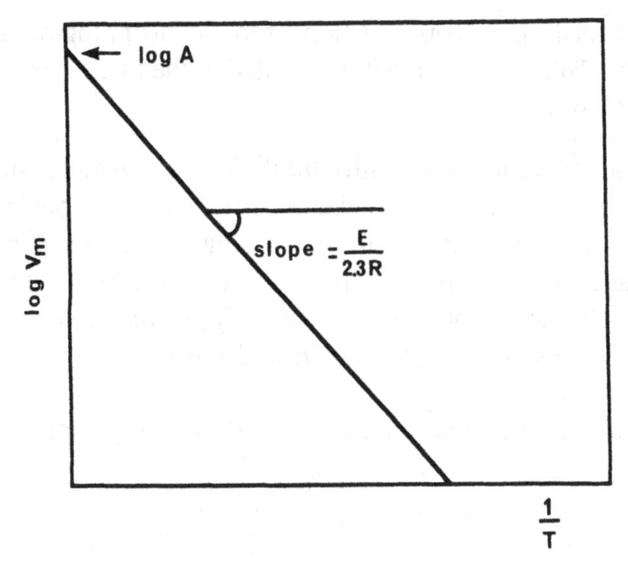

Figure 41. Arrhenius plot.

optimal temperature for the reaction. In some cases, changes of slope related to structural transition of an enzyme can occur.

1.1.4. Variation of the Reaction Rate as a Function of pH

The pH determines the ionization of some of the amino acid side chains in the enzyme, including those located in the active site. Protonation and deprotonation of these residues will affect the efficiency of substrate binding or catalysis. In some cases, changes in pH can affect the ionization of a group away from the active site but important for the correct folding of this site. The overall result of these phenomena is that most enzymes show an optimum pH of activity. When studying pH effects, it is important to take into account the possible involvement of ionizable groups belonging to the substrates.

1.1.5. Inhibition of Enzyme Activity

The rate of enzymatic reactions can be altered by numerous compounds. The influence of inhibitors that bind reversibly to the enzyme can be studied within the framework of the Michaelis theory. These phenomena are commonly used to obtain information regarding the mechanism of the enzymatic reactions (sequential or ping-pong mechanisms, etc.). This discussion will be limited to the three most common types of inhibition.

1.1.5a. Absolute Competitive Inhibition. Some molecules whose structure resembles that of the substrate may bind to the catalytic site of the enzyme in competition with the substrate. This type of inhibition does not affect V_m but alters K_M. This inhibition can be entirely reversed at high concentrations of substrate. The enzyme kinetics depend on the relative affinities of the inhibitor, I, and the substrate, S, and their concentrations.

This situation is described by the following equations:

$$E + S \rightleftarrows ES \rightleftarrows E + P$$
$$E + I \rightleftarrows EI$$

The dissociation constant of the inhibitor is

$$K_I = \frac{[E][I]}{[EI]}$$

and the rate equation is given by

$$V = \frac{V_m[S]}{K_M\left(1 + \frac{[I]}{K_I}\right) + [S]} \tag{42}$$

an expression that is similar to the Michaelis equation 36 with an apparent Michaelis constant

$$K'_M = K_M\left(1 + \frac{[I]}{K_I}\right)$$

According to Eadie (1952), the rate equation is

$$V = V_m - \frac{V}{[S]}K_M\left(1 + \frac{[I]}{K_I}\right) \tag{43}$$

The inverse of equation 42 is

$$\frac{1}{V} = \frac{K_M}{V_m}\left(1 + \frac{I}{K_I}\right)\frac{1}{[S]} + \frac{1}{V_m} \tag{44}$$

Graphs representating equations 42 and 44 are presented in Figs. 42 and 43, respectively.

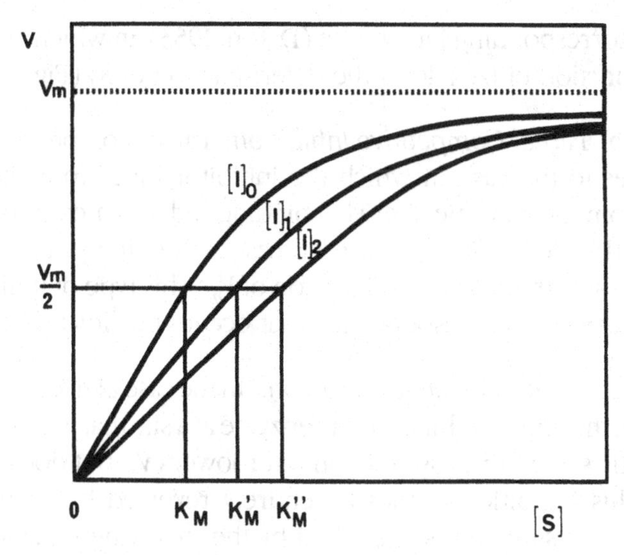

Figure 42. Plot of V as a function of $[S]$ in the case of absolute competitive inhibition. $[I]_0$ = no inhibitor; $[I]_2 > [I]_1$; K'_M and K''_M = apparent K_M.

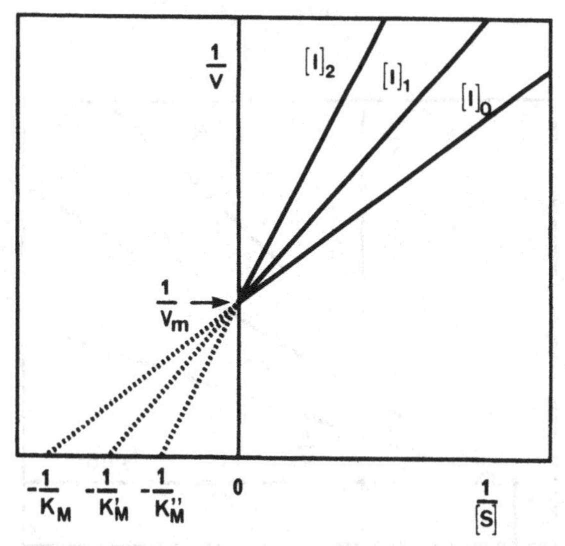

Figure 43. Lineweaver–Burk double-reciprocal plot in the case of absolute competitive inhibition. $[I]_0$ = no inhibitor; $[I]_2 > [I]_1$; K'_M and $K''M$ = apparent K_M.

The corresponding Dixon plot (Dixon, 1953), in which $1/V$ is plotted as a function of $[I]$, allows the determination of K_I (Fig. 44).

1.1.5b. Partial Competitive Inhibition. Partial competitive inhibition relates to the case in which the inhibitor binds to a site that is distinct from the catalytic site. This binding induces a conformational change in the protein, leading to a decrease of the affinity of the catalytic site for the substrate, without alteration of V_m. This type of inhibition is important, since it corresponds to what occurs in allosteric enzymes.

1.1.5c. Noncompetitive Inhibition. In the case of noncompetitive inhibition, the inhibitor binds to the enzyme at a site that is distinct from the catalytic site. In this case the inhibitor lowers V_m but does not alter K_M, and this inhibition cannot be entirely reversed by an excess of substrate. This situation is described by the following equations:

$$E + S \rightleftarrows ES \rightleftarrows E + P$$
$$E + I \rightleftarrows EI$$
$$ES + I \rightleftarrows ESI$$

In the simple theoretical case, ESI is inactive and cannot produce P, thus,

$$K_I = \frac{[E][I]}{[EI]} = \frac{[ES][I]}{[ESI]}$$

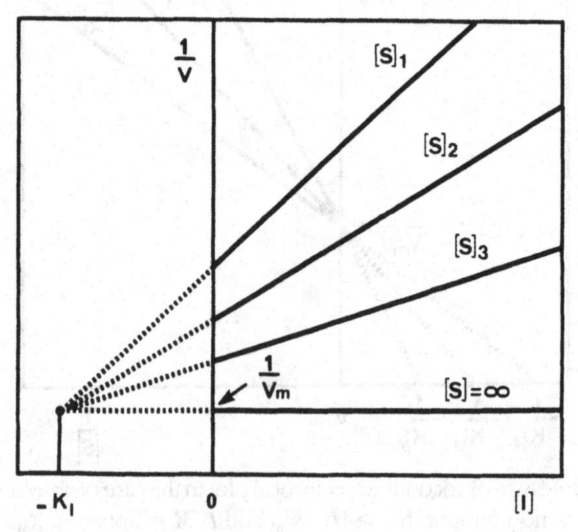

Figure 44. Dixon plot (Dixon, 1953). $[S]_3 > [S]_2 > [S]_1$.

The Michaelis rate equation becomes:

$$V = \frac{V_m}{\left(\frac{1 + K_M}{[S]}\right)\left(1 + \frac{[I]}{K_I}\right)} \tag{45}$$

whose inverse is

$$\frac{1}{V} = \left(1 + \frac{[I]}{K_I}\right)\left[\left(\frac{K_M}{V_m}\right)\left(\frac{1}{[S]}\right)\right] + \left(1 + \frac{[I]}{K_I}\right)\frac{1}{V_m} \tag{46}$$

The corresponding graphs are presented in Figs. 45 and 46.

The corresponding Dixon plot, $1/V = f/([I])$, allows the determination of K_I (Fig. 47).

Pure noncompetitive inhibition is rarely encountered, and most of the time a mixed type of inhibition is observed in which both K_M and V_m are altered.

1.1.6. Activation of Enzyme Activity

Contrary to what is described above, some compounds are able to stimulate the enzymatic activity by binding to specific sites. Numerous

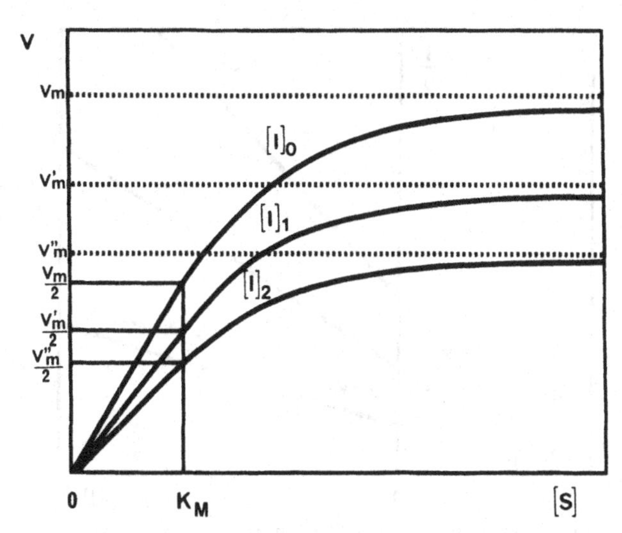

Figure 45. Noncompetitive inhibition; direct representation of V as a function of $[S]$. V'_m = apparent maximal velocity; $[I]_0$ = no inhibitor; $[I]_2 > [I]_1$.

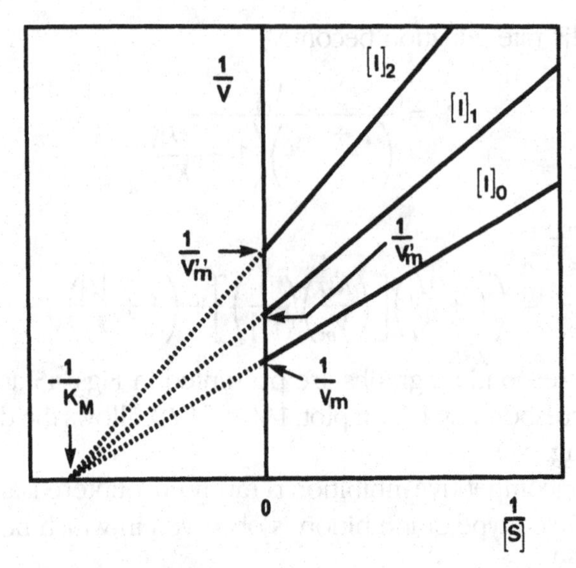

Figure 46. Noncompetitive inhibition; Lineweaver–Burk double-reciprocal plot. V'_m = Apparent maximal velocity; $[I]_0$ = no inhibitor; $[I]_2 > [I]_1$.

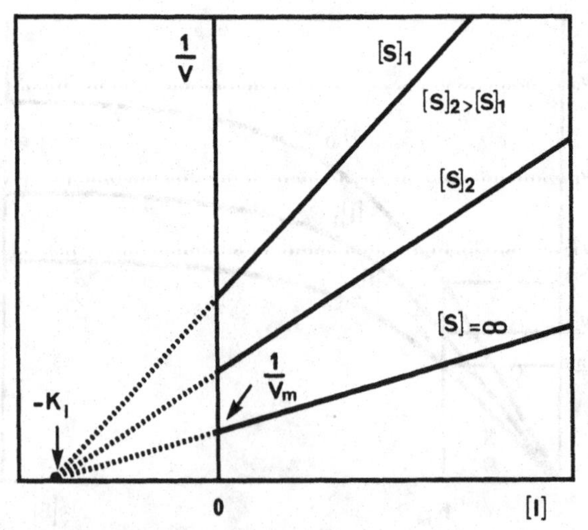

Figure 47. Noncompetitive inhibition; Dixon plot.

regulatory enzymes exhibit this property. The activator can increase the affinity for the substrate, the catalytic efficiency, or both. These various possibilities are represented in the following scheme:

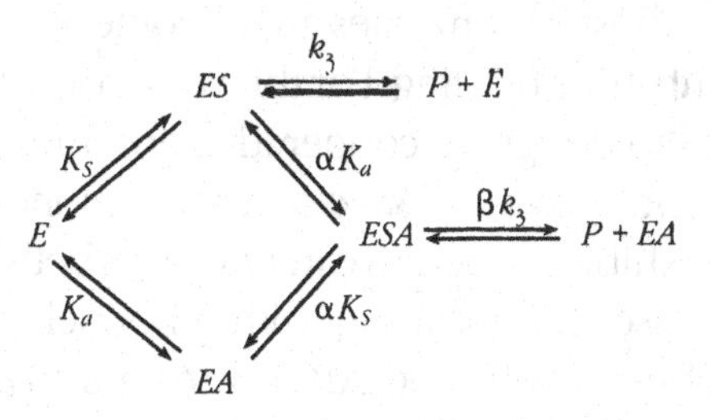

in which K_S is the dissociation constant of the substrate, S, that is, k_2/k_1 (equation 32), and K_a is the dissociation constant of the activator A. This general scheme covers all possible cases. The coefficients α and β account for the fact that the activator can act either on the dissociation constant of the substrate or on the catalytic constant. In terms of these parameters, the rate of reaction is given by

$$V = \frac{V_m[S_o](\alpha K_a + \beta[A])}{[S_o][A] + \alpha(K_a[S_o] + K_S[A] + K_S K_a)} \tag{47}$$

1.2. Allosteric and Cooperative Enzymes

If one considers cellular metabolism as an integrated set of a few thousand chemical reactions, it is easy to conceive the physiological necessity of regulating them. In the anabolic pathways, one frequently encounters the phenomenon of *feedback inhibition*, in which the final metabolite inhibits the activity of the first enzyme of the pathway that leads to its production:

$$A \underset{}{\overset{E_1}{\rightleftharpoons}} B \overset{E_2}{\rightleftharpoons} C \overset{E_3}{\rightleftharpoons} D \overset{E_4}{\rightleftharpoons} E \overset{E_5}{\rightleftharpoons} F$$

(−)

The enzymes that are subject to this type of regulation are called *allosteric enzymes* (Monod *et al.*, 1963). This term was coined to express the idea that the feedback inhibitor, F, binds to a specific site (*regulatory site*) distinct from the *catalytic site* that binds the substrate. This assumption has been confirmed thus far for all cases of feedback-inhibited enzymes for which structure of sufficient resolution has been determined. Similarly, but in the opposite sense, some enzymes are stimulated by *activators*, a phenomenon that is often involved in the cross-regulation of different metabolic pathways.

In most cases, allosteric enzymes are oligomeric and exhibit *cooperative effects* for substrate binding. For this reason, the term "allosteric" has also been used to designate cooperative enzymes. However, there are several allosteric enzymes (*stricto sensu*), monomeric or oligomeric, that do not exhibit cooperative interactions between the catalytic sites. This is the case, for instance, with ribonucleotide reductases (Eriksson and Sjöberg, 1989) and ATCase from *Saccharomyces cerevisiae* (Belkaïd *et al.*, 1987). The behavior of these enzymes corresponds to the phenomena of activation and inhibition of Michaelian enzymes (preceding section).

Positive cooperativity between the catalytic sites consists in the fact that the binding of the first molecules of substrate (or ligand) to an oligomeric protein facilitates the binding of the following ones. In this way, the affinity of the enzyme for its substrate increases with the level of occupancy of the binding sites. In such a case, the curve of the rate reaction V as a function of substrate concentration $[S]$ is no longer hyperbolic but sigmoidal (Fig. 48). The consequence of this is that the

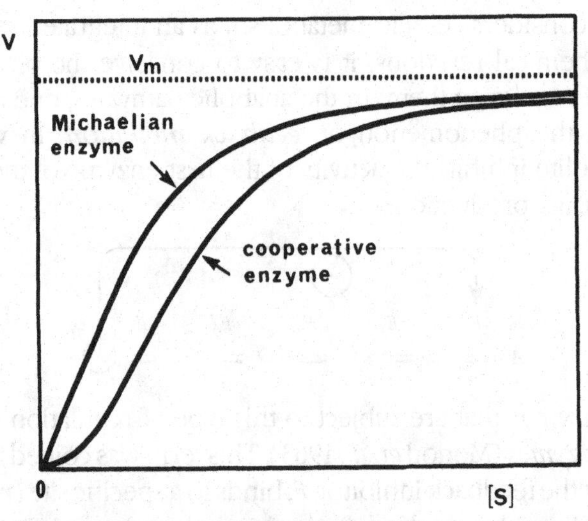

Figure 48. Substrate saturation curves of a Michaelian and a cooperative enzyme. The graph represents the simple case in which the cooperativity in substrate binding is not associated with a change in the catalytic constant.

substrate saturation curve cannot be linearized as previously described. Figure 49 shows the result obtained in Lineweaver–Burk and Eadie–Hofstee plots.

Some enzymes show negative cooperativity between the catalytic sites (Henis and Levitzki, 1989).

1.2.1. Theoretical Models for Cooperativity

1.2.1a. Hill Equation. The first theoretical analysis of protein cooperativity was proposed by Hill (1913). The formulation used does not make any assumption concerning the molecular mechanism involved. Consider an enzyme, E_n, made of n subunits, each binding one molecule of substrate

$$E_n + nS \rightleftarrows E_nS_n$$

with a binding constant

$$K = \frac{[E_nS_n]}{[E_n][S]^n} \qquad (48)$$

The Hill plot involves representing the variation of

$$\log \frac{[E_nS_n]}{[E_n]}$$

as a function of $\log[S]$. In practice, one takes $[E_nS_n]/[E_n]_{\text{total}}$ as the fraction, $\bar{Y}$, of sites occupied by the substrate, the fraction of nonoccupied sites being equal to $1 - \bar{Y}$. Writing equation 48 in this way gives

$$K = \frac{\bar{Y}}{[1 - \bar{Y}][S]^n}$$

and one obtains the ***Hill equation***

$$\log \frac{\bar{Y}}{1 - \bar{Y}} = \log K + n \log[S] \qquad (50)$$

In the case of an enzyme, and making the hypothesis that the reaction rate is simply proportional to the concentration of the complex ES, one actually plots the variation of $\log(V/V_m - V)$ as a function of $\log[S]$ (Fig. 50). The slope at the inflection point is called n_H, the *Hill coefficient*. Although the value of n_H is theoretically between 1 and

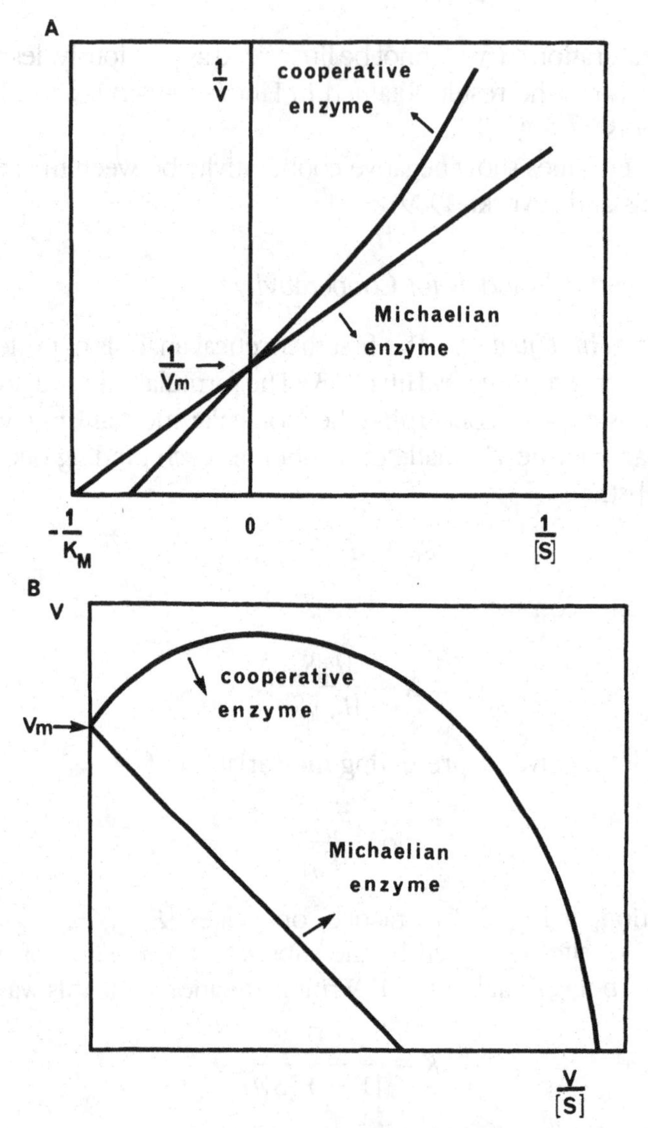

Figure 49. Substrate saturation curves of a Michaelian and a cooperative enzyme in the Lineweaver–Burk (A) and Eadie–Hofstee (B) plots.

n (the number of catalytic sites per molecule of oligomeric enzyme), the Hill plot is used mainly to provide a mathematical estimate of the degree of sigmoidicity of the saturation curve analyzed. For extreme substrate concentrations, n is equal to 1. With the graphical determinations shown in Fig. 50, one can determine the binding constants k_1 and k_n, characteristic of substrate binding to the first and nth sites, respectively.

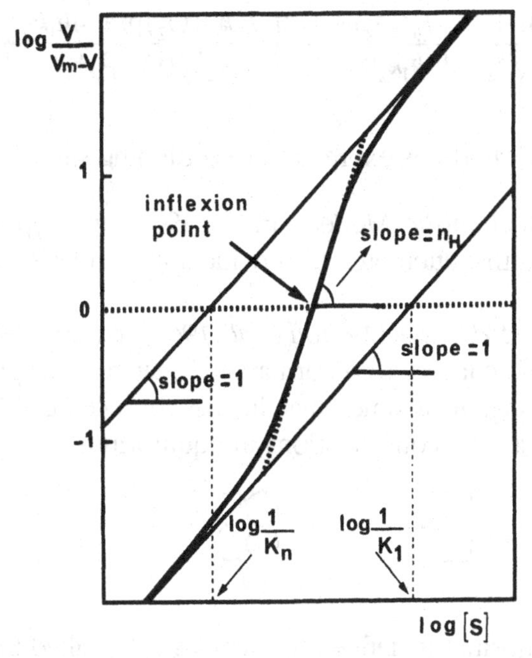

Figure 50. Hill plot of the substrate saturation curve of a cooperative enzyme.

1.2.1b. Adair Equation. In 1925, after showing that hemoglobin is a tetramer, Adair proposed to simply explain the sigmoidicity of its oxygen saturation curve in terms of four binding constants. This purely thermodynamic model does not make any assumption regarding the molecular mechanism

$$H_b + O_2 \overset{k_1}{\rightleftarrows} H_b(O_2) \qquad \text{with } k_1 = \frac{[H_b(O_2)]}{[H_b][(O_2)]}$$

$$H_b(O_2) + O_2 \overset{k_2}{\rightleftarrows} H_b(O_2)_2 \qquad \text{with } k_2 = \frac{[H_b(O_2)_2]}{[H_b(O_2)][(O_2)]}$$

$$H_b(O_2)_2 + O_2 \overset{k_3}{\rightleftarrows} H_b(O_2)_3 \qquad \text{with } k_3 = \frac{[H_b(O_2)_3]}{[H_b(O_2)_2][(O_2)]}$$

$$H_b(O_2)_3 + O_2 \overset{k_4}{\rightleftarrows} H_b(O_2)_4 \qquad \text{with } k_4 = \frac{[H_b(O_2)_4]}{[H_b(O_2)_3][(O_2)]}$$

On the basis of these constants, the fraction of sites $\overline{Y}$ occupied by oxygen is given by

$$\bar{Y} = \frac{k_1[(O_2)] + 2k_1k_2[(O_2)]^2 + 3k_1k_2k_3[(O_2)]^3 + 4k_1k_2k_3k_4[(O_2)]^4}{4(1 + k_1[(O_2)] + k_1k_2[(O_2)]^2 + k_1k_2k_3[(O_2)]^3 + k_1k_2k_3k_4[(O_2)]^4)} \quad (51)$$

One can extend this expression to n binding sites.

1.2.1c. Mechanistic Models. In the 1960s, two types of model were proposed that attempted to provide a molecular explanation of cooperativity.

The Concerted Model (Monod et al., 1965). This model postulates the existence of a concerted conformational change of the protein upon substrate binding. It assumes that in the absence of substrate, the enzyme exists as two conformations in equilibrium:

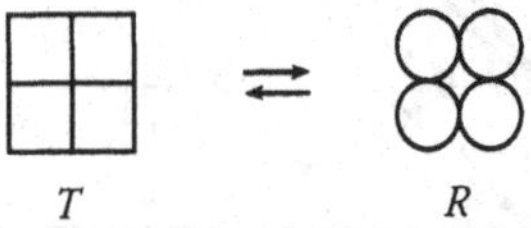

$$T \rightleftharpoons R \qquad L_o = \frac{[T]}{[R]}$$

This equilibrium is defined by a constant L_o, called the *allosteric constant*. By definition, the T and R forms have respectively a low and a high affinity for the substrate. An additional postulate of this model is that the transition between T and R is concerted, with no intermediary hybrid conformations. Binding of the substrate to the R form shifts the equilibrium toward this form. Within the framework of this model, the saturation function is:

$$\bar{Y} = \frac{L_o C\alpha(1 + C\alpha)^{n-1} + \alpha(1 + \alpha)^{n-1}}{L_o(1 + C\alpha)^n + (1 + \alpha)^n} \quad (52)$$

where n is the number of substrate binding sites per enzyme molecule and C is the ratio of the substrate dissociation constant for the two forms R and T. C is called the nonexclusion coefficient since it expresses the fact that the substrate does not bind exclusively to the R conformation, contrary to what would happen in a so-called exclusive system. α is $[S]/K_R$, the substrate concentration normalized to its dissociation constant K_R.

This simple model is frequently used in first attempts to interpret the behavior of cooperative enzymes. One of its characteristics is that it cannot explain the phenomenon of negative cooperativity, since the substrate cannot shift the $T \rightleftharpoons R$ equilibrium toward the conformation for which it has the lowest affinity. To overcome this defect, Viratelle and Seydoux (1975) proposed a modified version of the model.

The Sequential Model (Koshland et al., 1966). This more general model can account for the different types of cooperativity. Unlike the preceding model, which stresses the variations of quaternary structure, the sequential model puts more emphasis on the variations of the tertiary structure of the subunits and their interactions. The fundamental postulates of this model are:

1. In the absence of substrate or ligand, the protein exists as a single form called *A*.

2. When substrate binds to this form, it ***induces*** a transition of the subunit involved to another conformation called *B*.

3. By virtue of the interactions between subunits, this change induces a modification of the tertiary structure of the neighbor subunits. This phenomenon can have a positive or a negative influence on their affinity for the substrate, which thus binds preferentially to conformation *A* or *B*.

In this model, the saturation function is expressed in terms of the following parameters:

1. The constant of preferential binding to one of the two conformations:

$$K_s = \frac{[BS]}{[B][S]}$$

2. The constant of the transition from conformation *A* to conformation *B*:

$$K_t = [B]/[A]$$

3. The variations of the interactions between subunits, expressed by:

$$K_{AB} = \frac{(AB)(A)}{(AB)(B)}$$

$$K_{BB} = \frac{(BB)(A)(A)}{(AA)(B)(B)}$$

$$K_{AA} = 1 \qquad \text{by definition}$$

(*AB*) represents the interacting subunits, and (*A*) and (*B*) represent the noninteracting subunits. If one considers a tetramer, the nature of the interactions between subunits will depend on their spatial arrangement:

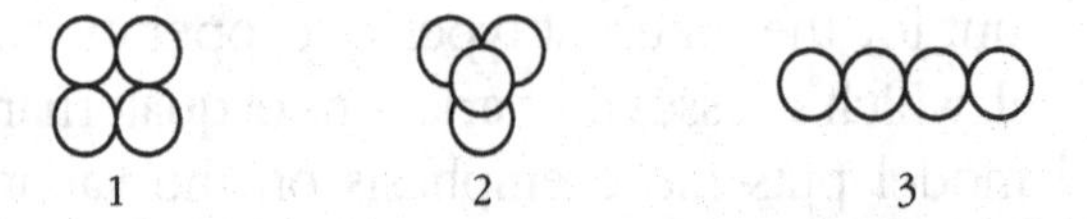

In case 1, for instance, the saturation function is given by:

$$\bar{Y} = \frac{K_{AB}^2[K_sK_t(S)] + (K_{AB}^4 + 2K_{AB}^2K_{BB})[K_sK_t(S)]^2}{1 + 4K_{AB}^2[K_sK_t(S)] + (2K_{AB}^4 + 4K_{AB}^2K_{BB})[K_sK_t(S)]^2} \cdots \frac{+\ 3K_{AB}^2K_{BB}^2[K_sK_t(S)]^3 + K_{BB}^4[K_sK_t(S)]^4}{+\ 4K_{AB}^2K_{BB}^2[K_sK_t(S)]^3 + K_{BB}^4[K_sK_t(S)]^4} \quad (53)$$

1.2.1d. Thermodynamic Coupling between Substrate Binding and Subunit Interactions. Weber (1972) analyzed the cooperative binding of a ligand, S, to an oligomeric protein strictly in terms of variations of free energy associated with ligand binding on one hand and interactions between subunits on the other.

If one considers the extreme case of a monomer–dimer equilibrium, the binding of ligand S is described by the following cycle:

$$\begin{array}{ccc} 2\text{Mon} + 2S & \overset{\Delta G_0}{\rightleftharpoons} & \text{Dim} + 2S \\ \Delta G' \updownarrow & & \updownarrow \Delta G \\ 2\text{Mon}\ 2S & \underset{\Delta G_2}{\rightleftharpoons} & \text{Dim}\ 2S \end{array}$$

In terms of thermodynamics, the various ΔG terms correspond to the variations of free energy associated with each of these reactions. Since the variation of free energy involved in the transition from one state to another is independent of the pathway, the sum of these variations around the cycle is equal to 0, and

$$\Delta G - \Delta G' = \Delta G_2 - \Delta G_0$$

In other words, for each difference in affinity of the ligand for the two states of the protein (monomer and dimer), there is an equal difference in the energy of interaction between the subunits in the liganded and unliganded forms of the protein.

The progressive binding of the ligand to the protein can be analyzed according to the thermodynamic scheme shown in Fig. 51.

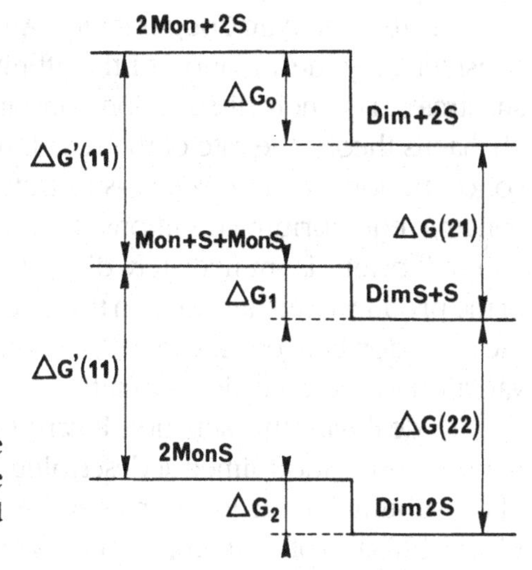

Figure 51. Relationship between the free energy of ligand binding and the association of the protomers. (Adapted from Weber, 1972, with permission.)

By definition, a *first-order* system is the one in which it is the binding of the first molecule of ligand, S, which changes the energy of interaction between the subunits. In the so-called *second-order* system, it is the binding of the second molecule of substrate which provokes this effect. The *intermediary order* refers to the case in which binding of the two molecules of ligands contributes. As a function of ΔG values, binding of the substrate will be either positively or negatively cooperative or noncooperative, and this binding will favor the formation or the dissociation of the dimer. The case presented in Fig. 51 is particular; it corresponds to a first-order system in which ligand binding is positively cooperative [$\Delta G(22) > \Delta G(21)$] and favors the dissociation of the protein into monomers [$\Delta G'(11) > \Delta G(21)$].

The same rationale applies to the case in which ligand binding is related not to complete dissociation of the protein but only to a variation of the energy of interaction between the subunits. From the different possible cases, it emerges that:

1. In a first-order system, ligand binding decreases the interaction between subunits if cooperativity is positive; it increases these interactions if cooperativity is negative.
2. In a second-order system, ligand binding increases the interaction between subunits if cooperativity is positive; it decreases this interaction if cooperativity is negative.

1.2.1e. Catalytic Cooperativity. All of the models discussed above consider only the variation of the affinity of the catalytic sites for the substrates, i.e., their dissociation constants. They imply that, as for the Michaelis theory, the rate of the reaction is simply proportional to the concentration of the enzyme–substrate complex. However, thermodynamic equilibrium conditions do not apply to enzymes, since such an equilibrium is immediately disrupted by catalysis and appearance of the products. Furthermore, in the case of some cooperative enzymes, the transition between the extreme conformations is accompanied by a variation of the catalytic constant.

To deal with this situation Ricard (1989) has developed a mathematical formulation aimed at describing the way in which the variations of the subunit interactions can alter the catalytic constant. According to Ricard, "What is of main importance is not to understand how subunits' interactions and conformational constraints modulate substrate binding to an enzyme, but in a more integrated way to understand how these interactions and constraints control the rate of product appearance under steady state conditions."

The rate constant of a reaction is related to its free energy of activation by the following equation:

$$k = \frac{K_B T}{h} \exp(-\Delta G^{\neq}/RT) \tag{54}$$

in which K_B is the Boltzman constant, h is the Planck constant, R is the universal gas constant, and T is the absolute temperature.

The basic idea of Ricard's model is that subunit interactions can have two different effects:

1. Subunit interactions can alter the rate of the conformational transitions involved in catalysis. This change of free energy of activation is called *promoter arrangement energy contribution*, $\Sigma({}^{\alpha}\Delta G^{\text{int}})$.
2. Subunit interactions can also cause a distortion of the catalytic site. This energetic contribution is called *quaternary constraint energy contribution*, $\Sigma({}^{\sigma}\Delta G^{\text{int}})$.

Therefore, the free energy of activation of the reaction will be

$$\Delta G^{\neq} = \Delta G^{\neq *} + \Sigma({}^{\alpha}\Delta G^{\text{int}}) + \Sigma({}^{\sigma}\Delta G^{\text{int}}) \tag{55}$$

in which $\Delta G^{\neq *}$ is called the *intrinsic energy contribution*. It corresponds to what the free energy of activation would be if the subunits were not interacting.

These energy contributions allow one to define a global velocity constant

$$k^{int} = \frac{K_B T}{h} \exp(-(\Delta G^{\neq *} + {}^{\alpha}\Delta G^{int} + {}^{\sigma}\Delta G^{int})/RT] \qquad (56)$$

The theoretical developments of this model led to interesting predictions, among which are the following:

1. Subunit interactions can induce catalytic cooperativity without sigmoidicity of the curve representing the variation of V as a function of $[S]$.
2. When the energy of subunit interactions is high, and if substrate binding is cooperative, catalytic cooperativity can only be positive.
3. Still considering the case in which the subunit interactions are strong, if substrate binding is negatively cooperative, catalytic cooperativity can be either positive or negative.

1.2.2. Influence of Effectors

The preceding discussion shows how complex the cooperativity for substrate binding or catalysis can be. These phenomena are frequently further complicated by the fact that many regulatory enzymes are also susceptible to the binding of regulatory metabolites (feedback inhibition, activation). In general, these effects alter the substrate saturation curve of the enzyme as shown in Fig. 52.

According to one of the postulates of the concerted model, it is generally assumed that the effectors are acting on the same equilibrium as the one involved in cooperativity between the catalytic sites. It appears that such is not the case for *E. coli* ATCase (Chapter 1). There are some indications that the PSE mechanism, which was proposed to account for the influence of ATP and CTP on the activity of this enzyme, could also explain the behavior of other regulatory enzymes (Hervé, 1989).

2. EXPERIMENTAL APPROACH

The enzymatic activity of ATCase and its isolated catalytic subunits is measured as the amount of carbamylaspartate (CAA) formed. Two alter-

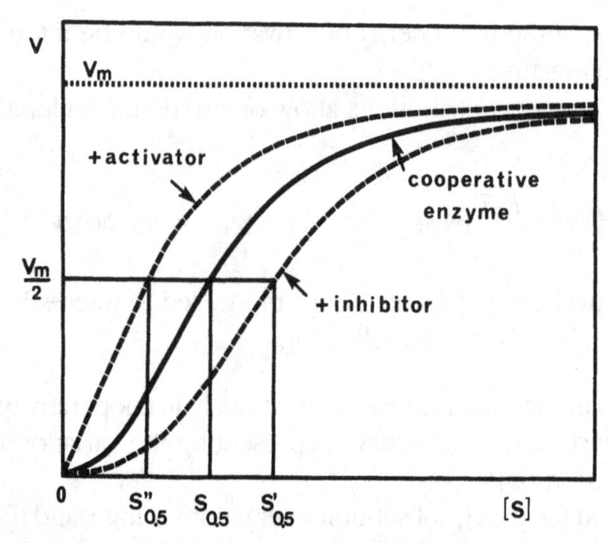

Figure 52. Influence of allosteric effectors on the substrate saturation curve of a positively cooperative enzyme. This scheme describes the case in which the effectors do not alter the catalytic constant of the enzyme. $S_{0.5}$ is the substrate concentration for which the reaction rate is equal to half the maximal velocity.

native methods are used to estimate the quantities of CAA: (1) titration of a radioactive precursor incorporated in the product and (2), a colorimetric assay.

2.1. Measurement of the Enzymatic Activity by Incorporation of [^{14}C]-Aspartate

2.1.1. Principle

One of the two substrates, aspartate (ASP), is labeled with ^{14}C (or ^{3}H). By measuring the amount of [^{14}C]-CAA formed, one can determine the enzymatic activity.

$$[^{14}C]\text{-ASP} + \text{carbamylphosphate} \xrightleftharpoons{\text{ATCase}} [^{14}C]\ \text{CAA} + \text{phosphate}$$

At the end of the reaction [^{14}C]-CAA is separated from [^{14}C]-ASP by ion exchange chromatography. The resin used (Dowex AG50W-X8) is made of polystyrene chains covalently linked by divinylbenzene groups. This polymer bears ionized sulfonate groups, $-SO_3^-$, which retain [^{14}C]-ASP, whose amino group is positively charged. On the other hand,

[14C]-CAA passes through the resin and is quantitatively estimated in the effluent by liquid scintillation.

2.1.2. Experimental Procedure

Equipment

- Water bath (37°C)
- Timer
- Pasteur pipettes and glass wool
- Liquid scintillation counter
- Test tubes and scintillation vials

Column Preparation

One Pasteur pipette is used for each test. Set up a series of these pipettes in a holder and push inside of each a small piece of wet glass wool as a stopper for the resin. Fill the pipette with a water suspension of the resin to obtain 4-cm-high columns (~0.8 ml of resin). Rinse the columns with 2 ml of distilled water and then place a counting vial under each. The columns must be used within 2 hr. (A device such as that shown in Fig. 53 is convenient.)

Solutions

- Reaction medium: 200-mM Tris–acetate buffer (pH 8.0) containing 20 mM carbamylphosphate (CAP), which should be dissolved in the buffer just before use
- 20 mM ASP (pH 8) of specific radioactivity approximately 0.15 mCi/mmole (or 0.5 mCi/mmole if [3H]-ASP is used)
- 2 mM CTP
- 20 mM ATP
- 0.2-N acetic acid
- Enzyme solutions: native ATCase (E), regulatory subunit (R; M_r = 33,940), reconstituted native ATCase (Erc) at 10 μg/ml, and catalytic subunits (C; M_r = 101,913) at 5 μg/ml. These solutions are prepared in the buffers described in Chapter 3 and in the Appendix. All of these solutions should be kept at 4°C

Minimum Precautions for the Use of Radioactive Compounds

- Wear special rubber gloves and rinse them after each contact with radioactivity
- Work on a bench covered with aluminum foil and absorbing filter paper

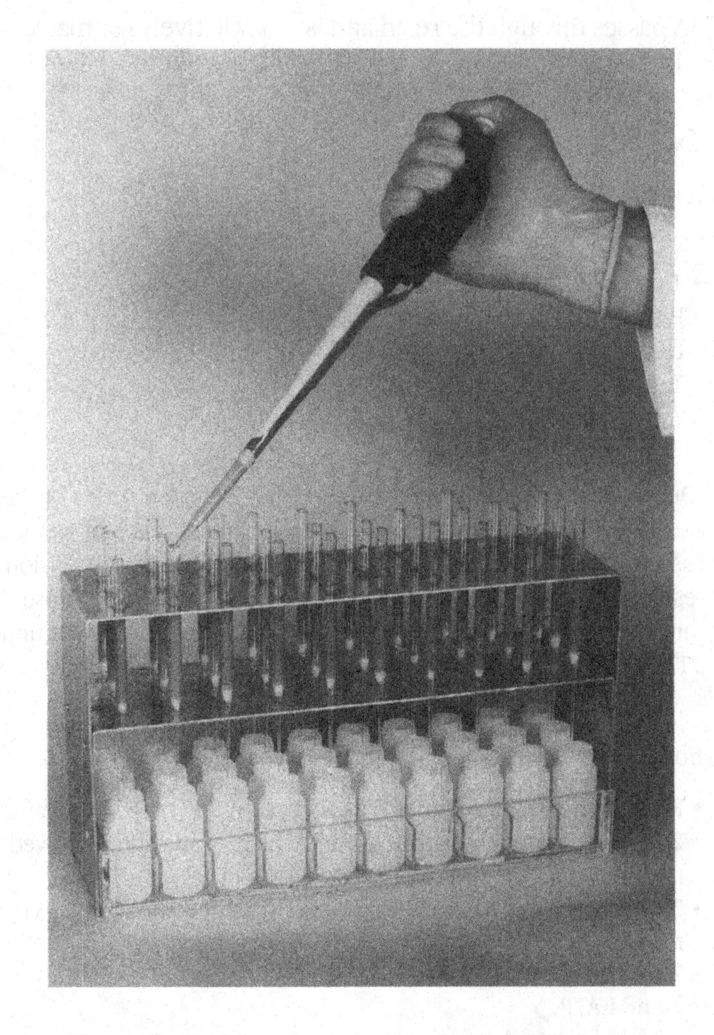

Figure 53. Useful equipment adapted to the ATCase assay.

- Never mouth pipette
- Follow laboratory instructions for the disposal of radioactive waste

Enzymatic Test

Set the incubator 37°C.

Prepare the incubation medium in test tubes at 0°C as indicated in Table 12 without adding the enzyme solutions. Tube 1 is a blank that will

Table 12. Procedure for the ATCase Assay Using [^{14}C]Aspartate

Volume (μl)	Tube number									
	1	2	3	4	5	6	7	8	9	10
Reaction medium	100	100	100	100	100	100	100	100	100	100
[^{14}C]aspartate	100	100	100	100	100	100	100	100	100	100
CTP	0	0	100	0	0	100	0	0	100	0
ATP	0	0	0	100	0	0	100	0	0	0
H_2O	200	100	0	0	100	0	0	100	0	100
Protein	0	100	100	100	100	100	100	100	100	100
			E			C		Erc		R

not receive enzyme but will be incubated and treated in the same way as the other samples.

Preincubate each test tube for 2 min at 37°C before starting the incubation by adding the enzyme (zero time). It is convenient to start the incubation of a sample every 30 sec. After 10 min of incubation, the reaction is stopped by the addition of 0.5 ml of 0.2-N acetic acid, which denatures the enzyme.

Estimation of the Amount of CAA Formed

Take a 0.5-ml sample from each test tube and place it on the top of one of the resin columns. When this sample has entirely entered into the resin, rinse the column four times with 0.5 ml of acetic acid, each time allowing the acid to enter entirely into the resin before making the next addition. To the 2.5 ml of eluate collected in the vial, add 8 ml of scintillation mixture (such as Aquasol, obtained from NEN). After agitation to homogenize, measure the radioactivity of the samples by using a scintillation counter. It is important to count a number of disintegrations high enough to ensure sufficient statistical precision. According to the Poisson distribution, a 95% confidence is given by $N + 2\sqrt{N}$, where N is the number of counts. Consequently, N must be large enough to ensure that $2\sqrt{N}$ is small compared to N. In general, about 10,000 counts will give satisfactory results ($2\sqrt{N}$ = 2%).

At the same time, determine the specific radioactivity of the [^{14}C]aspartate used by counting 20 μl of the 20-mM solution of this substrate in the presence of 2.5 ml of acetic acid and 8 ml of Aquasol.

If one uses a scintillation fluid that takes up less aqueous sample than does Aquasol, then an aliquot of the eluate is used for counting. For example, with 8 ml of Instagel (Packard), count 1 ml of the eluate of each column instead of 2.5 ml. Another possibility is to evaporate the sample to dryness and redissolve the residue in a small volume of distilled water.

Recycling of the Dowex AG50

After the experiments, it is possible to recycle the resin used by displacing the bound aspartate with 2-N HCl. This resin is then regenerated with 6-N HCl, 1-N NaOH, and 6-N HCl, successively. Between and after these treatments, the resin must be extensively washed with distilled water until the pH of the effluent is neutral. It is then resuspended and stored in distilled water. The [^{14}C]aspartate that has been eluted can be reused after purification.

Calculations and Presentation of Results

- Give precise information about the experimental conditions used.
- Express the results in counts per minute (cpm). Subtract the blank value (sample without enzyme) from each measurement.
- Express the experimental specific radioactivity of [^{14}C]aspartate in cpm per μmole and then express its specific radioactivity in microcuries per mmole, knowing that 1 Ci corresponds to 3.7×10^{10} disintegrations per second (dps), taking in account the efficiency of the counter that has been used (usually around 90% for ^{14}C and 30% for ^{3}H). What is the result in becquerels (see Appendix)?
- Calculate the amount of CAA formed in μmoles and express the specific activity of the different enzymatic preparations used in μmoles of CAA formed per hour per milligram of protein.

Question 11. *Does R have a catalytic activity? What are the effects of ATP and CTP?*

- Quantitate activation and inhibition as percentages of the control value as follows:

$$\text{Percent activation} = \frac{(V_A - V) \times 100}{V}$$

$$\text{Percent inhibition} = \frac{(V - V_I) \times 100}{V}$$

where V is the rate of reaction in absence of effector and V_A and V_I are the rates of reaction in the presence of activator and inhibitor, respectively.

- Was the recombination of C and R effective?

2.2. Measurement of Enzymatic Activity by Using a Colorimetric Assay

2.2.1. Principle

CAA, one of the products of the reaction, bears an ureido group (analog of urea)

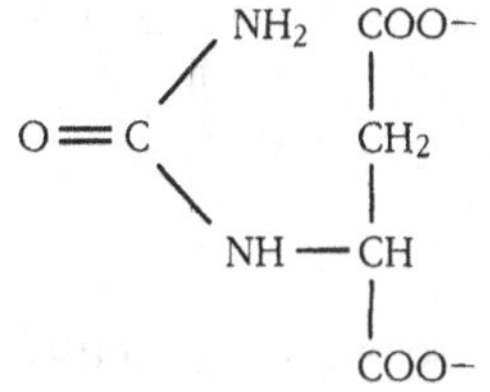

```
          NH2   COO-
         /       |
    O=C         CH2
         \       |
          NH —— CH
                 |
                COO-
```

This group can be condensed with antipyrine in the presence of sulfuric acid and diacetylmonoxime. The compound that is formed has a yellow–orange color (maximum absorption at 466 nm), which allows estimation of the amount of CAA formed (Prescott and Jones, 1969).

2.2.2. Experimental Procedure

Equipment

- Water bath at 37°C for enzyme incubation
- Covered water bath at 60°C for the colorimetric assay in the dark
- Spectrophotometer adjusted at 466 nm and plastic cuvettes

Reagents

- Antipyrine (1,5-dimethyl-2-phenyl-3-pyrazolone), 5 g in 1 liter of 50% (volume/volume) sulfuric acid

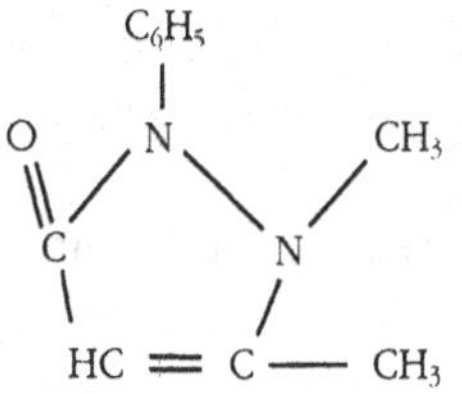

```
          C6H5
           |
    O      N         CH3
    ‖   /     \     /
    C           N
    |          /
    HC  ==  C —— CH3
```

Caution: Concentrated sulfuric acid is dangerous and can cause severe burning. Dissolve 5 g of antipyrine in 500 ml of distilled water, and to this solution slowly add 500 ml of pure sulfuric acid. Since mixing the acid with water results in a reaction that is highly exothermic, use Pyrex glassware and work over a sink in a fume cupboard. As a general rule, always add acid to H_2O, never H_2O to acid.

- Diacetylmonoxime (also called 2,3-butanedione monoxime). Dissolve 4 g of diacetylmonoxime in 500 ml of 5% (volume/volume) acetic acid

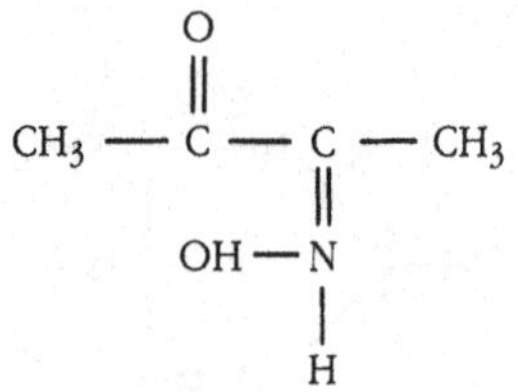

```
          O
          ||
CH3 — C — C — CH3
              ||
         OH — N
              |
              H
```

This solution is light sensitive and must be kept in a dark bottle at 4°C.

- Color mixture: This unstable mixture must be prepared just before use. It is made of one volume of the diacetylmonoxime solution, to which is slowly added two volumes of the sulfuric antipyrine reagent. The final solution must be kept in a dark bottle. It is then convenient to use a dispenser.
- Reaction mixture: 200-mM Tris–acetate buffer (pH 8) containing 20-mM CAP, which should be dissolved just before use
- 100-mM aspartate, pH 8
- 2-mM CTP and 20-mM ATP
- Enzyme solutions: E and R at 0.5 μg/ml; C at 0.25 μg/ml
- 1-mM CAA, pH 8

All of these solutions should be kept at 4°C.

Enzymatic Test

- Set the incubator at 37°C.
- Prepare the incubation media in test tubes at 4°C as indicated in Table 13 without adding the enzyme solutions. Tube 1 is a blank that will not receive any enzyme but will be incubated and treated like the other samples.
- Preincubate each test tube for 2 min at 37°C before starting the

Table 13. Procedure for the Determination of ATCase Activity by Using the Colorimetric Assay

	Tube number							
Volume (μl)	1	2	3	4	5	6	7	8
Reaction medium	250	250	250	250	250	250	250	250
Aspartate	50	50	50	50	50	50	50	50
CTP	0	0	250	0	0	250	0	0
ATP	0	0	0	250	0	0	250	0
H_2O	700	450	200	200	450	200	200	450
Proteins	0	250	250	250	250	250	250	250
			E			C		R

incubation by adding the enzyme (zero time). It is convenient to start the incubation of a sample every 30 sec. After 10 min of incubation, the reaction is stopped by adding 1 ml of the color mixture. Remember that this sulfuric solution can be dangerous; therefore, work in a hood. The tubes must be agitated, stoppered with marbles, and kept at 4°C in the dark until the next step.

Estimation of the Amount of CAA Formed

Incubate the samples at 60°C for 2.5–3 hr in the dark; after cooling the samples at room temperature, determine their absorbance at 466 nm against tube 1 (blank). The results are compared against a standard curve made with a series of ten test tubes containing increasing amounts of CAA (0–0.15 μmole) in 1 ml of water. Add 1 ml of the color mixture as described above.

Calculations and Presentation of Results

- Give precise information about the experimental conditions used.
- Draw the standard curve on graph paper.
- Calculate the amount of CAA formed in each test tube and express the specific activity of E and C in μmoles per hour per milligram of protein.
- How do these results compare with those obtained using the radioactive assay?

2.3. Enzymatic Properties of ATCase and Its Isolated Catalytic Subunits

The following experiments are designed to illustrate different basic aspects of enzyme studies. To perform these experiments, use the conditions described above except where indicated otherwise and use either the radioactive or the colorimetric assay. Do not forget to always include the blanks without enzyme.

Caution. To obtain good results, it is necessary to perform the experiments with great care and precision, especially in pipetting. All flasks and test tubes should be clearly labeled and arranged in an orderly fashion on a clean bench.

Equipment

As described before for both the radioactive (R.A.) and colorimetric (O.D.) assays.

Solutions

- Reaction medium: 200-mM Tris–acetate (pH 8) containing 20-mM CAP, which should be dissolved just before use.
- [^{3}H] or [^{14}C]aspartate solutions at concentrations of 20 mM and 80 mM. These solutions must be neutralized at pH 8 by NaOH. Their specific radioactivity must be adjusted as previously indicated (for R.A.).
- 100-mM unlabeled aspartate, pH 8 (for O.D.)
- 40-mM and 400-mM succinate, pH 7
- 2- 4- and 100-mM CTP, pH 7
- 40-mM UTP, pH 7
- 20-mM ATP, pH 7
- Enzyme solutions: E at 0.5 μg/ml (for O.D.) or 10 μg/ml (for R.A.); C at 0.25 μg/ml (for O.D.) or 5 μg/ml for (R.A.)

For pH dependence studies, a series of seven additional buffers must be prepared.

- 200-mM sodium cacodylate at pH 6.0, 7.0, and 7.5
- 200-mM Tris–acetate buffers at pH 7.5, 8.0, and 9.0
- 200-mM sodium–glycine buffer at pH 10.

CAP is added to each of these buffers just before use to a final concentration of 20 mM.

All of these solutions must be kept at 0°C and will be used to obtain the final experimental concentrations as indicated below. For each type of experiment, prepare a table describing the experimental conditions similar to Tables 12 and 13.

2.3.1. *Activity as a Function of Enzyme Concentration*

Experimental Conditions

- 50-mM Tris–acetate (pH 8.0)
- 5-mM CAP
- 5-mM [^{3}H] or [^{14}C]ASP
- Various amounts of enzymes (E or C), using stock solutions of 20 μg/ml
- Incubation time: 10 min at 37°C

Results (Fig. 54)

- Calculate the results and present them on graph paper.
- Up to what amount of product formed is the rate of reaction proportional to the enzyme concentration?

> Question 12. *What is the reason for the nonlinearity of the curve for the highest concentrations of enzyme?*

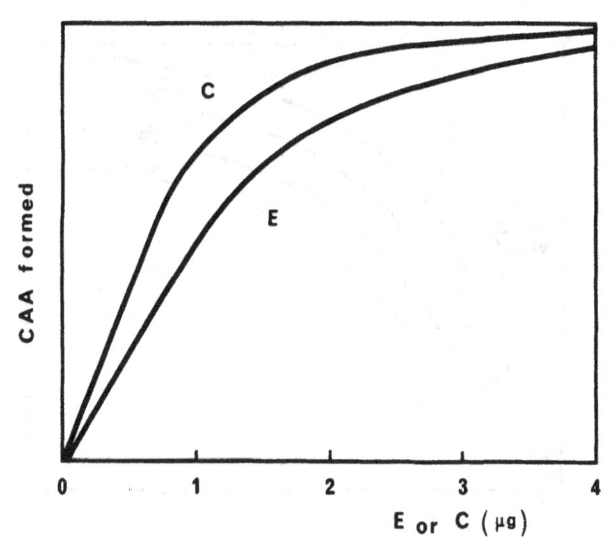

Figure 54. Rate of reaction as a function of the amount of enzyme.

Question 13. *Why is the rate of reaction higher with C than with E at equal concentration, and what is the ratio of these reaction rates in the linear part of the curve?*

Question 14. *What is the significance of the plateau that is reached in the presence of very high concentrations of enzymes?*

2.3.2. Kinetics of the Reaction at Different Aspartate Concentrations

Experimental Conditions

- 50-mM Tris–acetate, pH 8.0
- 5-mM CAP
- Enzyme solutions: E at 1 μg or C at 0.5 μg (for R.A.); E at 0.12 μg or C at 0.06 μg (for O.D.)
- ASP at 5, 10, 15, and 20 mM
- Various incubation times

Results (Fig. 55)

Question 15. *What is the information provided by this experiment?*

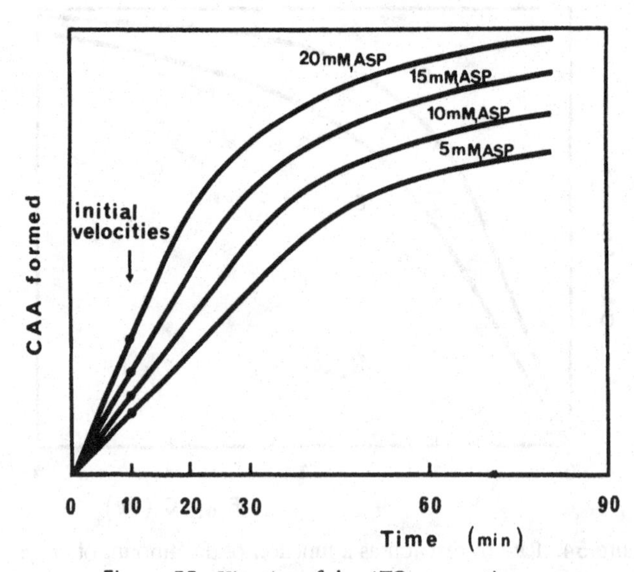

Figure 55. Kinetics of the ATCase reaction.

2.3.3. Rate of Reaction as a Function of Aspartate Concentration

Experimental Conditions

- 50-mM Tris–acetate, pH 8
- 5-mM CAP
- Enzyme solutions: E at 1 μg or C at 0.5 μg (for R.A.); E at 0.12 μg or C at 0.06 μg (for O.D.)
- ASP at various concentration from 0.5 to 40 mM, with several samples below 5 mM
- Incubation time: 5 min at 37°C

In the case of the radioactive assay, run a blank without enzyme for each ASP concentration. The values of the blanks will be subtracted from those of the corresponding samples.

Results (Fig. 56)

- Plot the variation of the blank as a function of ASP concentration and subtract each value from that of the corresponding enzymatic assay.
- Present the results obtained according to Michaelis and Menten, $V = f([S])$; Lineweaver and Burk, $1/V = f(1/[S])$; Eadie, $V = f(V/[S])$; Hanes, $([S]/V = f([S])$; Hill, $\log V/V_m - V = f(\log[S])$; and Eisenthal and Cornish-Bowden.

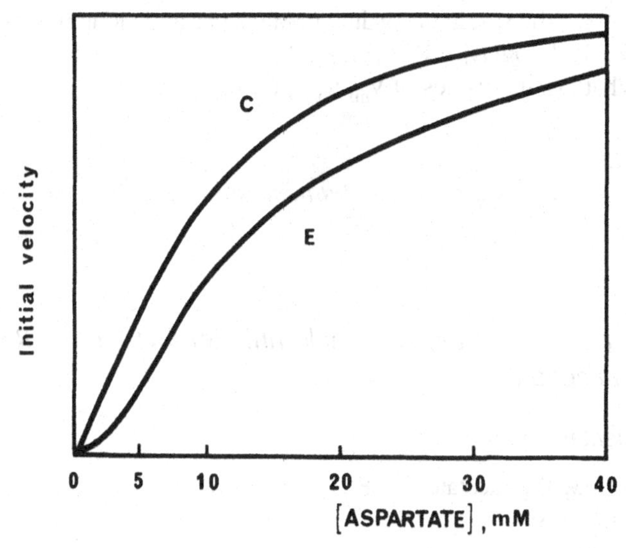

Figure 56. Aspartate saturation curve of ATCase (E) and its isolated catalytic subunits (C).

Question 16. *How do the saturation curves of C and E differ?*

- What are the values of V_m, K_M, $S_{0.5}$, *and* n_H for these two enzyme species?

2.3.4. Influence of the Effectors ATP and CTP on the Aspartate Saturation Curve of ATCase

Experimental Conditions

- 50-mM Tris–acetate, pH 8
- 5-mM CAP
- Enzyme solution: E at 1 μg (R.A.) or 0.12 μg (O.D.)
- ASP at various concentration from 0.5 to 40 mM, with several concentrations between 0.5 and 5 mM
- Incubation time: 5 min at 37°C
- In the case of the radioactive assay, run a blank without enzyme for each aspartate concentration. The values of these blanks will be subtracted from those of the corresponding samples.
- Establish an aspartate saturation curve in the presence and in the absence of 1-mM CTP and 5-mM ATP.

Results (Fig. 57)

- Present the results according to Michaelis and Menten, $V = f([S])$; and Hill, $\log V/V_m - V = f(\log[S])$.
- What are the values of V_m, $S_{0.5}$, and n_H?

Question 17. *What are the influences of the effectors ATP and CTP?*

2.3.5. Influence of the Feedback Inhibitor CTP on the Rate of Reaction

Experimental Conditions

- 50-mM Tris–acetate, pH 8
- 5-mM CAP
- 5-mM ASP

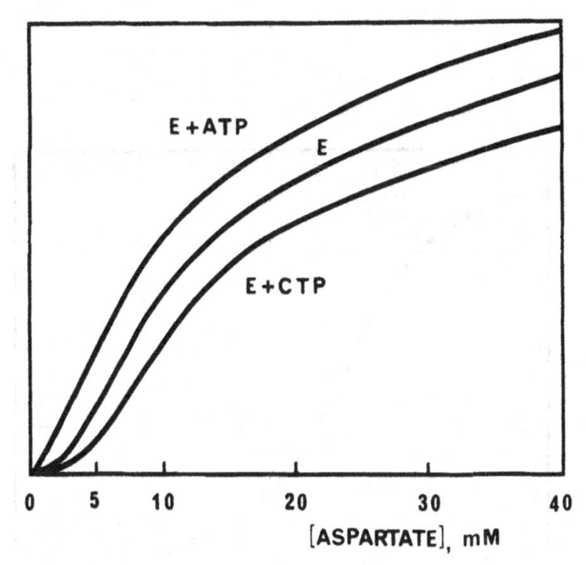

Figure 57. Influence of the effectors ATP and CTP on the aspartate saturation curve of ATCase.

- Increasing concentrations of CTP from 5 μM to 1 mM; with at least two controls run without CTP
- Enzyme solution: 1 μg (R.A.) or 0.12 μg (O.D.)
- Incubation time: 10 min at 37°C

Results (Fig. 58)

- Represent the rate of reaction and percent inhibition against CTP concentration on semi-logarithmic graph paper (use the formula given in Section 2.1).

2.3.6. Synergistic Effects of CTP and UTP

Experimental Conditions

- 50-mM Tris–acetate, pH 8
- 5-mM CAP
- 5-mM ASP
- Enzyme solution: 1 μg (R.A.) or 0.12 μg (O.D.)
- Incubation time: 10 min at 37°C
- Various CTP and UTP concentrations from 0 to 4 mM. In an additional series of samples containing CTP, add increasing concentrations of UTP from 2 to 4 mM.

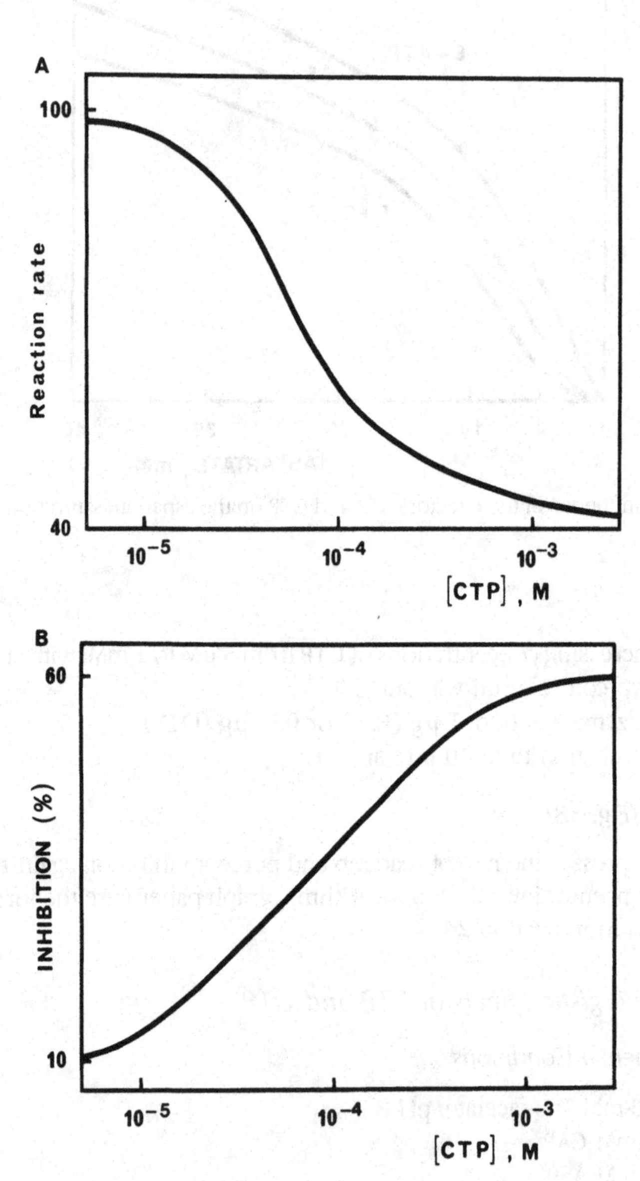

Figure 58. Influence of CTP on ATCase rate of reaction. (A) Influence on the rate of reaction; (B) variation of the percent inhibition.

Results (Fig. 59)

Plot the rate of reaction as a function of the CTP and UTP concentrations.

> Question 18. *What is the significance of the result obtained? Propose a molecular explanation.*

2.3.7. Influence of the Activator ATP on the Rate of Reaction

Experimental Conditions

- 50-mM Tris–acetate, pH 8
- 5-mM CAP
- 5-mM ASP
- Increasing concentrations of ATP from 10 μM to 10 mM, with at least two controls run without ATP
- Enzyme solution: 0.5 μg (R.A.) or 0.06 μg (O.D.)
- Incubation time: 10 min at 37°C

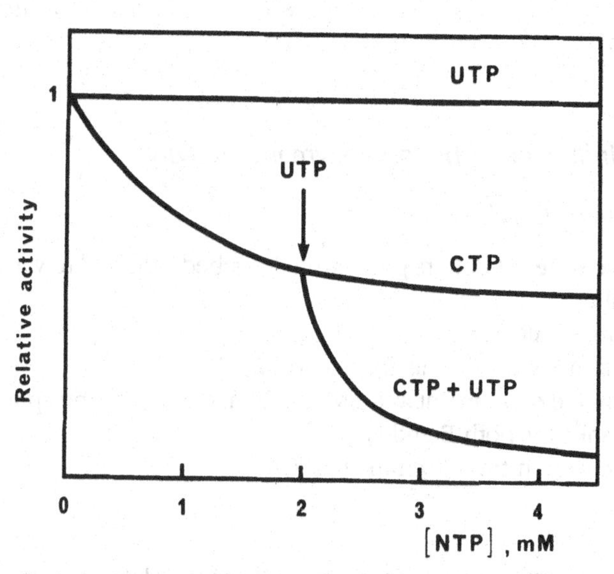

Figure 59. Influence of UTP, CTP, and their mixture on the rate of the ATCase reaction. (From Wild *et al.*, 1989, with permission of J. Wild.)

Results (Fig. 60)

Represent the rate of reaction and percent activation against ATP concentration on semilogarithmic graph paper (use the formula given in Section 2.1).

2.3.8. Competition between Inhibitor and Activator

Experimental Conditions

- 50-mM Tris–acetate, pH 8; 5-mM CAP and 1-mM ASP
- Enzyme solution: 1 μg (R.A.) or 0.12 μg (O.D.) of E
- 1-mM ATP and increasing concentrations of CTP from 10 μM to 30 mM
- At least two controls run without CTP
- Incubation time: 10 min at 37°C

Results (Fig. 61)

Present the stimulation or inhibition observed according to the formula given in Section 2.1.

Question 19. *What happens in the presence of both effectors at the same time, and what is the significance of the point at which there is neither activation nor inhibition?*

2.3.9. Influence of pH on Enzymatic Activity

Experimental Conditions

- The series of buffers previously described, adjusted to various pH values
- 5-mM CAP
- Enzyme solutions: as for Section 2.3.3.
- Run the experiment at 1-mM and 20-mM ASP with the appropriate blanks for both E and C
- Incubation time: 10 min at 37°C

Results (Fig. 62)

Plot the rate of reaction against pH for the different cases investigated.

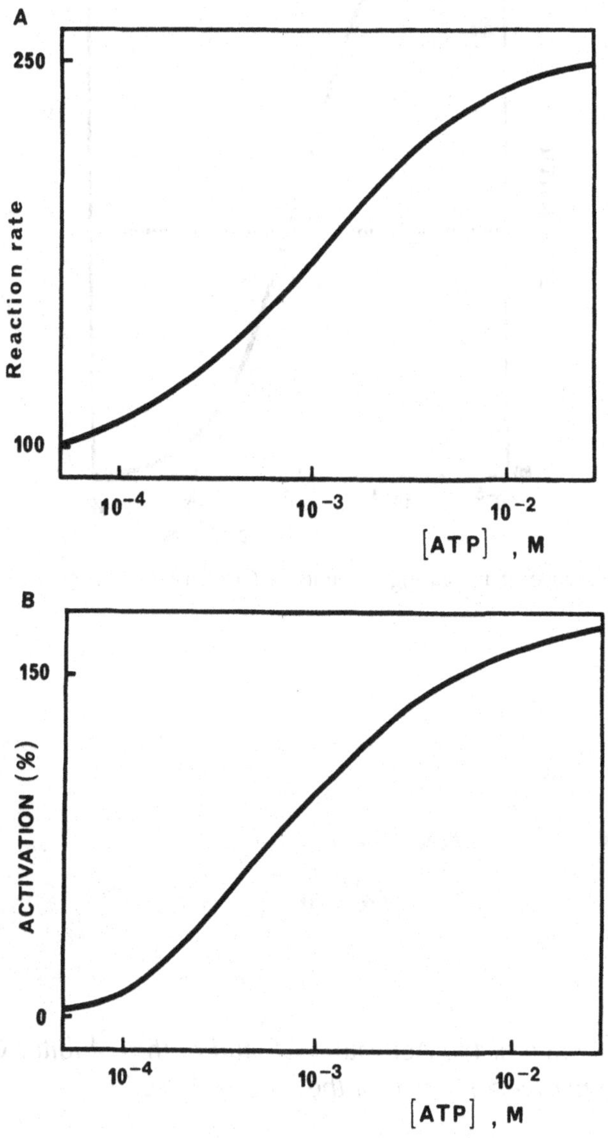

Figure 60. Influence of ATP on the rate of the ATCase reaction. (A) Reaction rate; (B) percent activation.

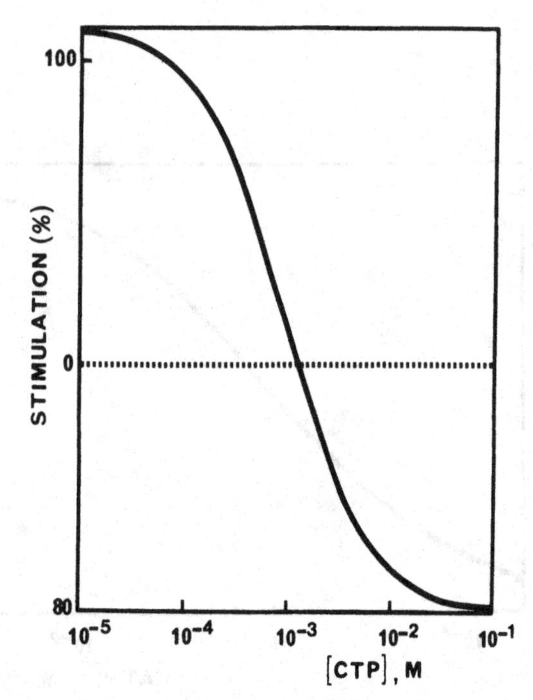

Figure 61. Influence of increasing amounts of CTP on the reaction rate of ATCase stimulated by ATP.

Question 20. *What are the optimum pH values? In what respect are E and C behaving differently? What is the meaning of this difference? What happens to E between 1- and 20-mM ASP (see experiment described in Section 2.3.3)?*

2.3.10. Influence of the Activator ATP and of the Inhibitor CTP on the pH Dependence of the ATCase Reaction

Experimental Conditions

- 50-mM Tris–acetate, pH 8; 5-mM CAP and 5-mM ASP
- Incubation time: 10 min at 37°C
- Absence or presence of 5-mM ATP
- Absence or presence of 5-mM CTP
- Enzyme solution: 1 μg (R.A.) or 0.12 μg (O.D.)

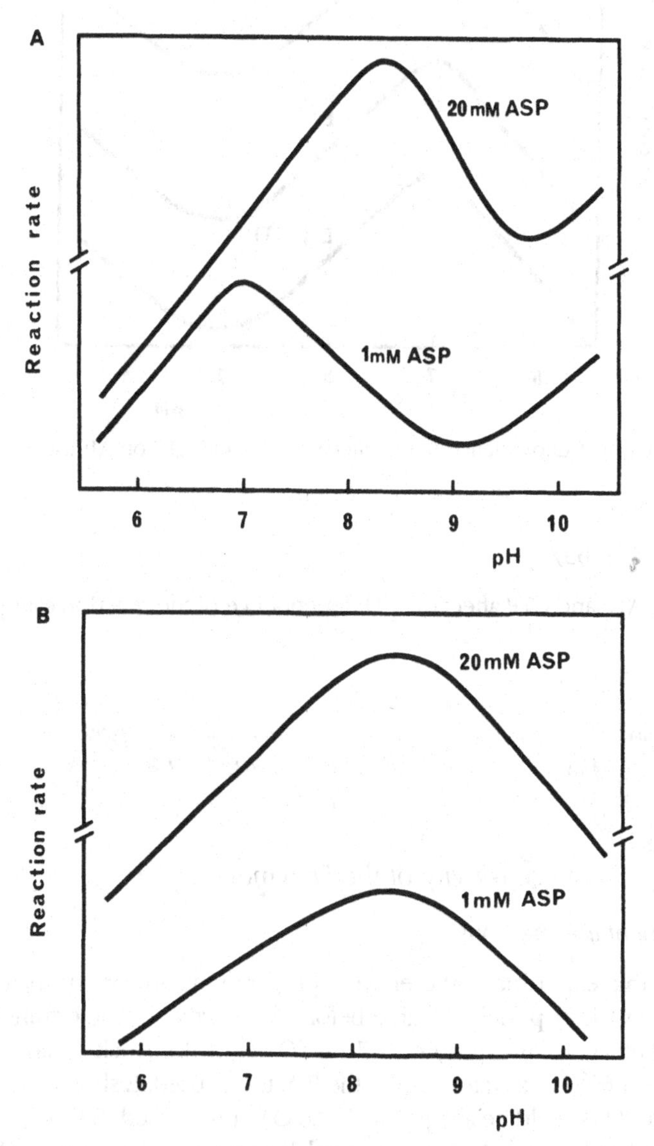

Figure 62. Influence of pH on the enzymatic activity of native ATCase (A) and its isolated catalytic subunits (B).

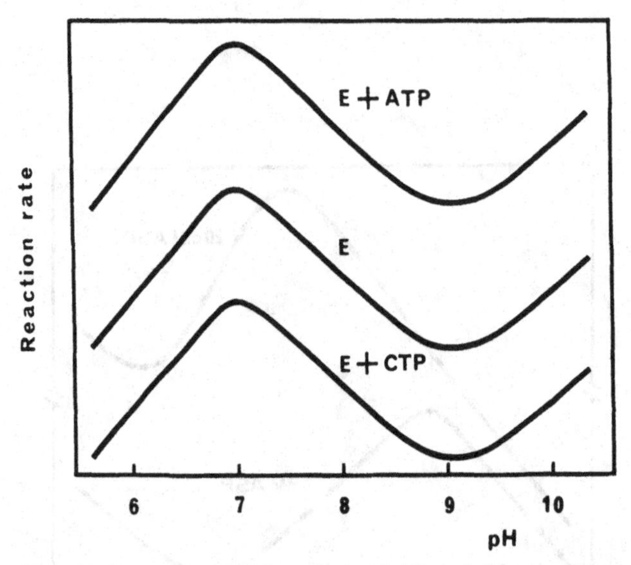

Figure 63. pH dependence of the effects of ATP and CTP on ATCase activity.

Results (Fig. 63)

Do ATP and CTP affect the pH dependence of the reaction catalyzed by E?

Question 21. *In view of the conclusions of the experiment described in Section 2.3.9, what is the meaning of these results?*

2.3.11. Thermosensitivity of the Enzymes

Experimental Conditions

In this experiment, the enzyme preparations are preincubated at 60°C for various periods of time before their activity is determined.

Incubate 1.2 ml (R.A.) or 2.7 ml (O.D.) of the stock solutions of enzyme at 60°C in a water bath. If the R.A. test is used, E should be at 10 μg/ml and C should be at 5 μg/ml. If the O.D. test is used, E should be at 0.5 μg/ml and C should be at 0.25 μg/ml. Remove samples of 120 μl (R.A.) or 270 μl (O.D.) at zero time and after 15 min, 30 min, 45 min, 1 hr, 1.5 hr, 2.5 hr, and 3 hr. Immediately cool these samples in ice water. Use 100 μl (R.A.) or 250 μl (O.D.) of each of these samples to determine their enzymatic activity by incubation for 10 min at 37°C in the presence of 50-mM Tris–acetate (pH 8), 5-mM CAP, and 20-mM ASP.

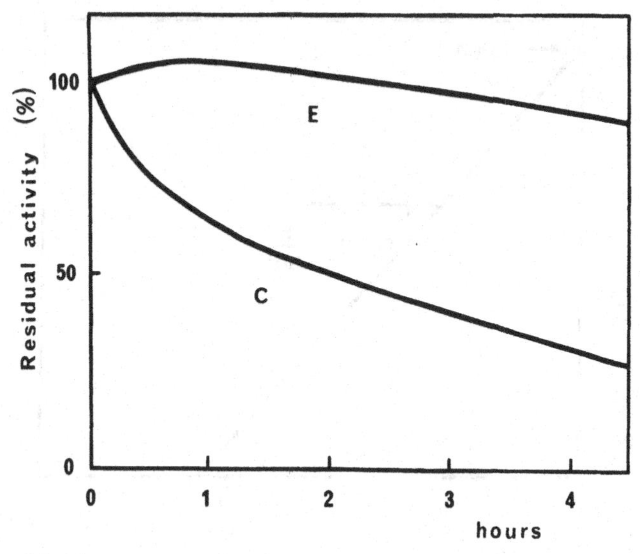

Figure 64. Thermosensitivity of ATCase and its isolated catalytic subunits.

Results (Fig. 64)

Plot the residual activity as a function of duration of exposure to 60°C.

Question 22. *What is the interpretation of the result obtained?*

2.3.12. *Influence of Temperature on the Enzymatic Activity of the Catalytic Subunits*

Experimental Conditions

- Same as for the experiment described in Section 2.3.3
- Incubations at various temperatures from 5 to 40°C, with 5°C steps
- Double and triple the C concentrations in the test for incubation at the lowest temperatures

Results (Fig. 65)

- Plot the results obtained as $\log V_M$ against the reciprocal of the absolute temperature (Arrhenius plot).
- Calculate the free energy of activation of the reaction.

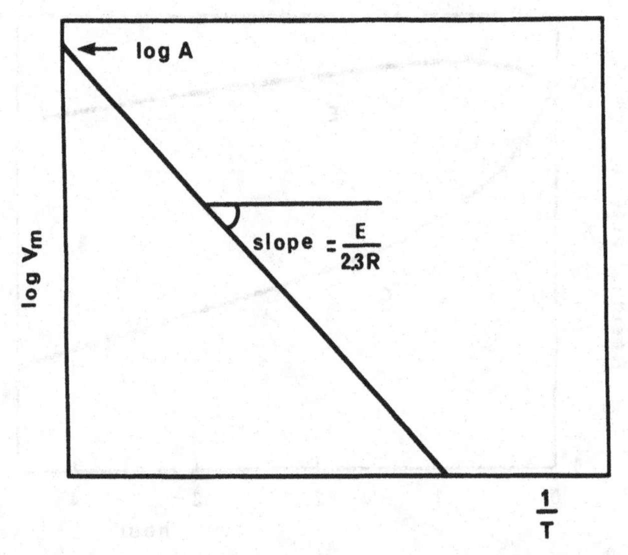

Figure 65. Influence of temperature on the reaction catalyzed by ATCase (Arrhenius plot).

2.3.13. *Influence of Succinate on ATCase Activity*

2.3.13a. *Inhibition of the Catalytic Subunits by Succinate*

Experimental Conditions

- 50-mM cacodylate buffer, pH 7; 5-mM CAP and various ASP concentrations as in Section 2.3.3
- Absence and presence of 1-, 5-, and 50-mM succinate, pH 7
- Enzyme solution: C at 0.5 μg (R.A.) or 0.06 μg (O.D.)
- Incubation time: 10 min at 37°C

Results (Fig. 66)

Plot $V = f[\text{ASP}]$ at the different succinate concentrations used and the corresponding double-reciprocal plots. Determine the values of K_M and V_m.

Question 23. *What kind of inhibitor is succinate?*

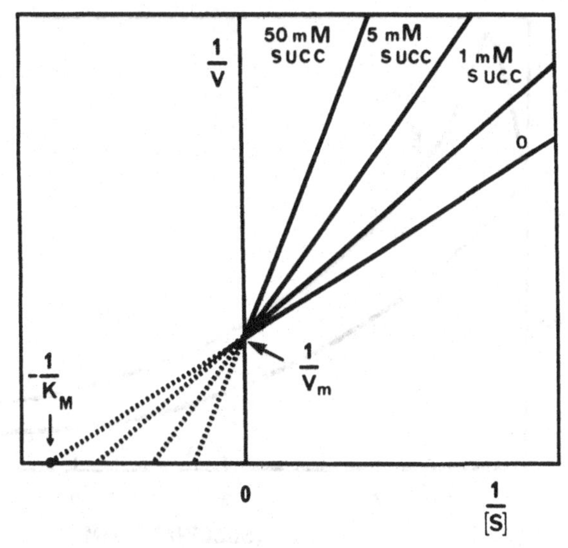

Figure 66. Lineweaver–Burk plot of the aspartate saturation curve of the catalytic subunits in the presence of succinate.

Calculate the K_I of succinate from the formula that accounts for the inhibition, where

$$\frac{K_M}{V_m}\left(1 + \frac{[I]}{K_I}\right)$$

is the slope of the curves in the double-reciprocal plot (see equation 44).

Compare the value obtained by drawing the corresponding Dixon plot, $1/V = f([I])$.

2.3.13b. *Influence of Succinate on Native ATCase*

Experimental Conditions

- 50-mM cacodylate buffer, pH 7; 5-mM CAP and 1-mM ASP
- Incubation time: 10 min at 37°C
- Absence and presence of succinate at concentrations of 0.2 to 10 mM
- Enzyme solutions: as a control, C is also tested under the same conditions. E and C are used at the concentrations indicated in Section 2.3.3

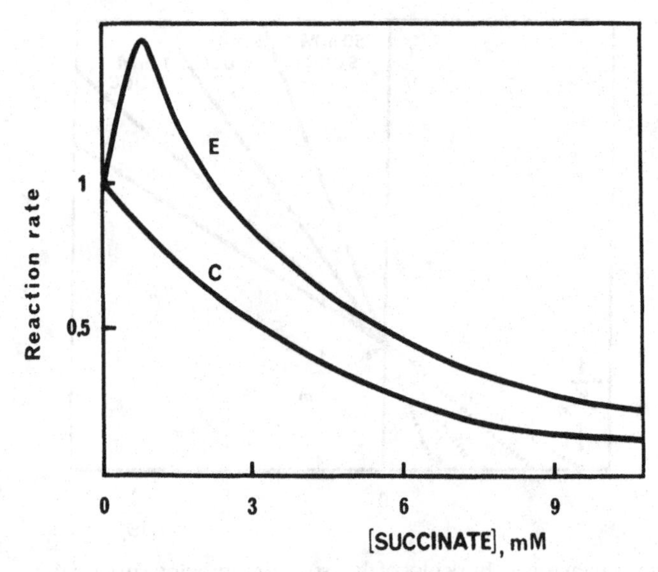

Figure 67. Influence of succinate on the activity of ATCase and its isolated catalytic subunits.

Results (Fig. 67)

Plot *V* as a function of succinate concentration for E and C, *V* being expressed in relative units.

Question 24. *In what ways do E and C behave differently under these conditions? Knowing the allosteric properties of E, what is the explanation for the increase in the rate of reaction that is observed at low succinate concentrations?*

Chapter 6

Complementary Experiments

The experimental program presented in the preceding chapters is not exhaustive. ATCase allows the elaboration of numerous other experiments of educational interest. The following are some possibilities.

1. MOLECULAR BIOLOGY

1. Isolation of the plasmid that contains the ATCase operon; determination of its restriction map and DNA sequencing.
2. Use of this highly expressed plasmid for analysis of the mRNA coding for ATCase.

2. PHYSICAL CHEMISTRY

1. Difference spectroscopy: determination of the binding isotherms of substrate analogs (cooperative binding of succinate in the presence of carbamylphosphate, for example). Because of its very high affinity for the catalytic site of ATCase and its isolated catalytic subunit, the bisubstrate analog *N*-phosphonacetyl-L-aspartate allows the direct titration of these catalytic sites (Collins and Stark, 1971; Kerbiriou *et al.*, 1977).
2. Binding of the effectors ATP, CTP, and UTP to the regulatory sites by equilibrium dialysis; continuous-flow dialysis or airfuge ultracentrifugation.
3. Determination of the tyrosine and tryptophan content of the protein by the method of Bencze and Schmidt (1957).

4. After slab gel electrophoresis, transfer to an Immobilon membrane and immunological detection (Towbin *et al.*, 1979).
5. Determination of the sedimentation coefficients in sucrose gradients (Smith, 1988).
6. Sedimentation equilibrium using a preparative ultracentrifuge (Minton, 1989).
7. HPLC size exclusion chromatography (le Maire *et al.*, 1986).

3. ENZYME KINETICS

1. Computer processing of the kinetic results.
2. Analysis of the pH dependence for activity; determination of the pK_as of the chemical groups involved in substrate binding and catalysis (Léger and Hervé, 1988).
3. Test of the ordered mechanisms, using the isolated catalytic subunits.

Appendix

1. EQUATIONS AND UNITS

Although it is recommended that the International System of Units (SI in meters, kilograms, and seconds) be used whenever possible, the cgs system is still extensively used in biochemistry. For instance, concentrations and partial specific volumes are generally expressed in g/liter and cm^3/g, respectively, instead of kg/m^3 and m^3/kg. Furthermore, the expression of molar mass is commonly in g/mol and not kg/mol (Edsall, 1970). This unit derives directly from the use of the cgs system in equations 15 and 17 (see below). As far as teaching is concerned, it is most important to prepare students to use and convert the different units encountered in the literature.

The cgs system converts to SI as follows:

1. force: 1 dyne ($g{\cdot}cm{\cdot}s^{-2}$) = 10^{-5} newton ($kg{\cdot}m{\cdot}s^{-2}$)
2. energy: 1 erg ($g{\cdot}cm^2{\cdot}s^{-2}$) = 10^{-7} joule ($kg{\cdot}m^2{\cdot}s^{-2}$) = 2.39×10^{-8} cal

where g = gram and s = second.

The units used for some of the equations are indicated below.

1. Equation 2: $f = 6\pi\eta R$; equation 16: $f = 6\pi\eta R_S$
   f = frictional coefficient in $g{\cdot}s^{-1}$
   η = viscosity in poise ($dyne{\cdot}cm^{-2}{\cdot}$ or $g{\cdot}s^{-1}{\cdot}cm^{-1}$)
   R or R_S: radius or Stokes radius in cm
2. Equation 3: $\mathbf{v} = Q\mathbf{E}/f$
   $\mathbf{v}$ = rate in $cm{\cdot}s^{-1}$
   $Q\mathbf{E} = \mathbf{F}$: force in dynes
3. Equation 7: $d(\ln r_M) = \omega^2 s dt$

ω = angular velocity in rad·s^{-1}

s = sedimentation coefficient in s·rad^{-2} (usually expressed in seconds, s)

A sedimentation coefficient of 1×10^{-13} s·rad^{-2} is called a Svedberg (1 S). Therefore, a sedimentation coefficient of 8×10^{-13} s·rad^{-2} is 8 S.

4. Equation 13: $D^0 = RT/Nf$

D^0 = diffusion coefficient in cm^2·s^{-1}

$R = 83.14 \times 16^6$ erg·°K·mol^{-1}

$N = 6.023 \times 10^{23}$ mol^{-1}

5. Equation 15: $M = \dfrac{(d\ \ln c/dr^2)2RT}{(1 - \bar{v}\rho)\omega^2}$

Molar mass in g·mol^{-1}

$$\text{g·mol}^{-1} = \frac{\text{cm}^{-2} \times 2(\text{g·cm}^{2}\text{·s}^{-2}\text{·°K}^{-1}\text{·mol}^{-1})\text{°K}}{[1 - (\text{cm}^3\text{·g}^{-1})(\text{g·cm}^{-3})]s^{-2}}$$

Example: The molar mass of ATCase as determined by sedimentation equilibrium is 310,000 g/mol.

In the SI system one has, with r, the distance to the axis of rotation expressed in meters and R, the gas constant = 8.3 J·mol^{-1}·°K^{-1}:

$$\text{kg mol}^{-1} = \frac{\text{m}^{-2} \times 2(\text{kg·m}^{2}\text{·s}^{-2}\text{·°K}^{-1}\text{·mol}^{-1})\text{·°K}}{[1 - (\text{m}^3\text{·kg}^{-1})(\text{kg·m}^{-3})]s^{-2}}$$

Thus, in the SI system the molar mass of ATCase becomes 310 kg/mol.

6. Equation 17: $M = \dfrac{6\pi\eta R_s s_{20,w} N}{1 - \bar{v}\rho}$

$$\text{g·mol}^{-1} = \frac{6\pi(\text{g·s}^{-1}\text{·cm}^{-1})\text{cm·s·mol}^{-1}}{[1 - (\text{cm}^3\text{·g}^{-1})(\text{g·cm}^{-3})]}$$

2. MOLAR MASS AND MOLECULAR MASS

Equations 15 and 17 allow the determination of M, the molar mass in g·mol^{-1}, which is numerically equivalent to the molecular mass expressed in daltons (1 dalton = 1.661×10^{-24} g = 1/12 of the mass of one atom of nuclide ^{12}C).

The difference between these two concepts arises from the fact that one considers in the first case 6.023×10^{23} identical molecules (1 mol) and in the second case a single molecule whose mass is expressed in terms of a multiple of that of the nuclide ^{12}C atom.

One may recall that Avogadro's number arises from the fact that 6.023×10^{23} atoms of nuclide ^{12}C (mass: $12 \times 1.661 \times 10^{-24}$ g) make 12 g of ^{12}C.

Example: ATCase has a molecular mass of $305{,}646 \times 1.661 \times 10^{-24}$ g $= 5.07678 \times 10^{-19}$ g·molecule^{-1} or $305{,}646/(5.07678 \times 10^{-19}) = 6.023 \times 10^{23}$ molecules per mol (Avogadro's number).

In the case of complex molecules (ribosomes, viruses, chromosomes, mitochondria, etc.): it is suggested that one use the mass in daltons, rather than the molar mass expressed in g·mol^{-1} or the *relative molecular mass* (M_r), which refers to the mass of the ^{12}C atom (Edsall, 1970). The relative molecular mass is dimensionless.

It is correct to say either that the molar mass of X is 10,000 g·mol^{-1} or that its molecular mass is 10,000 daltons (10 kdaltons) or that its relative molecular mass is 10,000. It is incorrect to say that the molecular *weight* of X is 10,000 daltons.

Physical Meaning of Mass and Weight

It is important to clearly distinguish the concepts of weight (gravitational force exerted on an object by the earth) and mass (amount of matter in the object). Two instruments illustrate the difference between these two notions, the dynanometer and the balance. In both cases, one uses a solid as a reference. Through the deformation of its spring, the dynanometer measures the weight of a given amount of matter, which is a force whose exact value depends on the place on earth where the measurement is made. Therefore, it does not determine an unambiguous property of the matter. The balance makes use of an equilibrium between antagonistic couples. It provides absolute results independent of the place where the measurement is made. Therefore, it provides a nonambiguous expression of the amount of matter, the *mass*.

The ultracentrifuge allows one to determine the molecular or molar mass but not the molecular or molar weight.

3. UNITS OF CATALYTIC ACTIVITY

It is important to quantitate the catalytic efficiency of an enzyme. The enzymatic preparations are more or less purified and thus, in terms of units per milligram of protein, more or less active. The unit of activity

(U) is the amount of enzyme catalyzing the transformation of 1 μmole of substrate per minute under standard conditions. The following conditions must be precisely defined: ionic strength, pH, temperature (often 25 or 37°C), etc. The standard conditions are chosen by the experimentalist and defined for each enzyme. It is preferable to use a "saturating" concentration of substrate or the optimum concentration when inhibition by excess substrate is observed. In the case of ATCase, the activity cannot be measured at saturating aspartate concentrations since a concentration corresponding to ten times the K_M (that is, 0.2 M) could provoke an inhibition due to the high ionic strength of the medium, a phenomenon that can be observed in the presence of KCl. Moreover, it is known that, like many other enzymes, ATCase is inhibited by an excess of substrate.

When the molecular mass of the enzyme is not known, if the protein is not pure, or if its activity is measured under nonoptimal conditions, one can define a *specific activity*, that is, the amount of substrate transformed per unit of time per milligram of enzyme under precisely defined standard conditions: $\mu mol \cdot min^{-1} \cdot mg^{-1}$ or $U \cdot mg^{-1}$.

When the molecular mass of a pure enzyme is known, one defines its *molecular activity* (*turnover number*), which is the number of substrate molecules transformed by one molecule of enzyme per second at optimal substrate concentration and under standard conditions: s^{-1}.

It may be recalled that except for particular purposes, enzymatic activity must be measured under conditions ensuring a constant rate of reaction (initial velocity). These conditions are met when the substrate concentration used is saturating and as long as the concentrations of substrate and product are far from those defined by the equilibrium constant of the reaction. In experiments in which substrate is not saturating, the rate of reaction must be measured under conditions in which only a small proportion of substrate is transformed (less than 10%).

The Nomenclature Committee of the International Union of Biochemistry (1979) has recommended the use of the *katal* (abbreviation: kat) to express the catalytic activity of enzymes. The katal is the catalytic activity that will raise the rate of reaction by 1 mol/s in a specified assay system. In this way, the specific activity can be expressed in $kat \cdot kg^{-1}$ instead of $\mu mol \cdot min^{-1} \cdot mg$ of $enzyme^{-1}$

$1\ \mu mol \cdot min^{-1} \cdot mg^{-1} = 1\ U \cdot mg^{-1} = 16.67\ mkat \cdot kg^{-1}$

4. UNITS OF RADIOACTIVITY

The unit of radioactivity is now the *becquerel* (abbreviation: Bq).
1 Bq = one disintegration per second
1 Curie (Ci) = 3.7×10^{10} disintegrations per second
Therefore, 1 Ci = 3.7×10^{10} Bq = 37 GBq.

5. UNITS OF QUANTITY

The *quantities* of substances are expressed in mole, mmole, μmole, etc. or, in recent publications (SI system) in mol, mmol, μmol, etc. By contrast, *concentrations* are expressed as follows:

Solutions	Quantity	Usual symbol	SI
Molar	1 mole per liter	M	$mol{\cdot}l^{-1}$
Millimolar	1 millimole per liter	mM	$mol{\cdot}l^{-1}$
Micromolar	1 micromole per liter	μM	$\mu mol{\cdot}l^{-1}$

6. *k* FACTOR OF THE ULTRACENTRIFUGE ROTOR

k factors given in rotor manuals provide the time (in hours) that a molecule of 1 S will take to migrate from the top of the tube to its bottom at 20°C in water at the maximum allowed speed of the rotor.

Example a: For the SW41 rotor (maximum speed = 41,000 rpm); *k* = 124, a molecule of sedimentation coefficient 1 S will take 124 hr to reach the bottom of the tube, and a 10 S molecule will take 12.4 hr.

Example b: If the centrifugation is not performed at the maximum allowed speed of the rotor, the correction must take into account the squares of the speeds: for the SW41 rotor spinning at 20,000 rpm, to reach the bottom of the tube a 1 S molecule would take:

$$124 \times \frac{(41)^2}{(20)^2} = 521 \text{ hr}$$

A 10S molecule would take 52.1 hr.

7. CALCULATION OF ACCELERATION

The RCF, or *relative centrifugational force* to which a solution is exposed in a centrifuge tube, is calculated as follows:

$$RCF = \omega^2 r/g$$

where $g = 981\ \text{cm}\cdot\text{s}^{-2}$ and r is the distance from the axis of rotation (in centimeters). In general, one takes the bottom of the tube as the reference in order to calculate the maximum acceleration.

Example: For the SS34 rotor at 12,500 rpm, the acceleration at the bottom of the tube (i.e., at 10.8 cm from the rotational axis) is:

$$\frac{(2\pi\cdot 12{,}500)^2 \times 10.8}{(60)^2 \times 981} = 18{,}800\ \text{g}$$

8. BACTERIAL STRAINS

The bacterial strains described in Chapter 2, with which all of the proposed experiments can be carried out, can be obtained from G. Hervé, Laboratoire d'Enzymologie, CNRS, 91198 Gir-sur-Yvette, France.

9. SOLUTIONS AND REAGENTS

Some solutions, in particular those used for the gel electrophoresis experiments, are described in the corresponding chapters. The others are described here.

9.1. 0.8-M Tris–Acetate Buffer, pH 8

Tris ($M_r = 121$):	96.9 g
Adjust the pH at 8, using CH_3COOH	
Water:	up to 1 liter

9.2. 0.2-M Acetic Acid

CH_3COOH at 100%:	57.5 ml
Water:	up to 5 liters

9.3. 2×10^{-3}-M CTP

Na_3 CTP (M_r = 567):	113.4 mg
Adjust at pH 7–8	
Water:	up to 100 ml

9.4. 4×10^{-3}-M CTP

Na_3 CTP:	226.8 mg
Adjust at pH 7–8	
Water:	up to 100 ml

9.5. 2×10^{-2}-M ATP

Na_2 ATP (M_r = 551):	1.10 g
Adjust at pH 7–8	
Water:	up to 100 ml

9.6 Elution Buffer for Chromatography: 10× Stock Solution

TES (M_r = 229):	22.9 g
KCl (M_r = 74.6):	74.6 g
Azide (M_r = 65):	0.65 g
Adjust at pH 7.5	
Water:	up to 1 liter

9.7. 100-mM Aspartate, pH 8

L-Aspartic acid (M_r = 133.1):	13.31 g
Adjust at pH 8 (NaOH)	
Water:	up to 1 liter

9.8. 10-mM Carbamylaspartate, pH 8

Carbamylaspartic acid (M_r = 176):	1.76 g
Adjust at pH 8 (NaOH)	
Water:	up to 1 liter

9.9. 1-M Phosphate Buffer, pH 7.2

KH_2PO_4:	68 g
Adjust at pH 7.2 (KOH)	
Water:	up to 500 ml

9.10. Buffer for Dilution of E

0.8-M Tris–acetate, pH 8:	50 ml
14-M β-mercaptoethanol:	150 μl
Water:	up to 1 liter

9.11. Buffer for Dilution of C

1-M potassium phosphate, pH 7.2:	40 ml
100-mM EDTA:	2 ml
β-Mercaptoethanol:	150 μl
Water:	up to 1 liter

9.12. Buffer for Dilution of R and Recombination

0.8-M Tris–acetate, pH 8:	50 ml
β-Mercaptoethanol:	150 μl
Zn acetate·$2H_2O$ (M_r = 219.5):	22 mg
Water:	up to 1 liter

9.13. 200-mM Cacodylate Buffer, pH 6, 7, and 7.5

Sodium cacodylate·$3H_2O$ (M_r = 214):	8.56 g
Adjust at desired pH	
Water:	up to 200 ml

9.14. 200-mM Tris–Acetate Buffer, pH 8 and 9

Tris:	4.8g
Adjust at pH 8 or 9 (CH_3COOH)	
Water:	up to 200 ml

9.15. 200-mM Glycine Buffer, pH 10

Glycine (M_r = 75):	3 g
Adjust at pH 10 (NaOH)	
Water:	up to 200 ml

9.16. 400-mM Succinate, pH 7

Sodium succinate·6H_2O (M_r = 270):	54.03 g
Adjust at pH 7	
Water:	up to 500 ml

9.17. Reaction Medium: 20-mM CAP–200-mM Tris–Acetate, pH 8

Lithium carbamylphosphate (M_r = 152.9):	152.9 mg
0.8-M Tris–acetate, pH 8:	12.5 ml
Water:	up to 50 ml

10. PREPARATION OF STANDARD PROTEIN MIXTURES FOR COLUMN CALIBRATION

The amounts are given for pure proteins. In some cases, it is necessary to take into account the weight of the salts present. For the Sepharose calibration, one uses two proteins at a time.

Thyroglobulin (Pharmacia calibration kit)	2.5–3 mg
Aldolase (Pharmacia calibration) kit	2.5–3 mg
β-Galactosidase (Sigma, G6008)	3 mg
Cytochrome *c* (Sigma, C2506)	2.5 mg
Ferritin (Pharmacia calibration, kit)	1 mg
Ribonuclease A (Sigma, R5000)	5 mg
Bovine serum Albumin (Sigma, A7511)	5 mg
Cytochrome *c* (Sigma, C2506)	2.5 mg

Thyroglobulin (Pharmacia, calibration kit)	2.5–3 mg
Ovalbumin (Sigma, A2512)	5 mg
Catalase (Pharmacia calibration, kit)	2.5–3 mg
Myoglobin (Sigma, M0630)	2 mg
Transferin (Sigma, T2252)	3 mg
Hemoglobin (Sigma, H7379)	2 mg

These mixtures must be kept at 4°C in a desiccator.

Answers to Questions

Question 1. Use the linear part of the ATCase biosynthesis curve (Fig. 14, t = 10 min). The enzymatic test allows determination of the number of micromoles of carbamylaspartate formed per milliliter of extract per hour. Divide by 3000 (specific activity of ATCase under the conditions used) to obtain the amount (in milligrams per milliliter) of ATCase present at time t_1. Make the same calculation for time t_2 (for example, t = 12 min in Fig. 14). By difference, and dividing by $t_2 - t_1$ (in minutes), one obtains the amount (in milligrams) of ATCase synthesized in 1 ml of culture. Multiply by Avogadro's number and divide by the molecular mass of ATCase (M_r = 305,646) and by the number of bacteria per milliliter to obtain the number of ATCase molecules synthesized per bacterium per minute.

Question 2. The degradation of the RNA is monitored on the basis of the disappearance of its ability to direct the biosynthesis of ATCase. Determination of its half-life in the semilogarithmic plot (Fig. 16B) relies on the postulate that its inactivation is an exponential phenomenon.

Question 3. ATCase, 1.8×10^5 M^{-1} cm^{-1}; catalytic subunit (Cat), 0.73×10^5 M^{-1} cm^{-1}; regulatory subunit (Reg), 0.12×10^5 M^{-1} cm^{-1}.

Question 4. The Lowry reaction is very complex, and absorption of the compounds formed does not follow the Beer-Lambert law (equation 1). Consequently, the standard curve obtained is not a straight line, and the most probable curve is drawn through the experimental points.

Question 5. The ATCase sample shows an intense main band and two minor bands located on both sides. The slow-migrating band is a trace of the dimeric form of the enzyme present in all solutions. The fast-migrating band is an altered form of ATCase that lacks one regulatory dimer (C_6R_4). The enzymatic properties of the first are indistinguishable from those of the native enzyme. The regulatory properties of C_6R_4 are slightly diminished (Evans *et al.*, 1975; Subramani and Schachman, 1980). Concentrated solutions of ATCase always show this slight heterogeneity. The regulatory subunits appear as a diffuse band; in solution, the isolated regulatory subunits exist as monomers and dimers in an equilibrium which is shifted to the dimeric form in the presence of Zn. The monomer tends also to form nonspecific aggregates. Even though the catalytic subunits are higher in molecular mass than the regulatory subunits, their electrophoretic migration is faster because of their higher net charge at the pH used.

Question 6. For determination of the stoichiometry of the two types of chains in the native enzyme, two prerequisites must be fulfilled.

1. The two kinds of chains must bind the Coomassie blue equally on a mass basis. Example: For 3 μg of ATCase deposited on top of the gel, one has 2 μg of the 34,000-dalton chain and 1 μg of the 17,000-dalton chain. If this condition is met, the heavy chain will produce a band of intensity twice that of the light chain.
2. The intensity of the coloration must be measured under conditions in which the Beer-Lambert law is valid; it is necessary to verify that there is proportionality between the optical density and the amount of protein deposited on the gel. This condition must be met for both types of chains.

Experimentation shows that when the band coloration is measured with a densitometer under condition in which requirement 2 is ensured, one has: optical density of the 34,000-dalton chain/optical density of the 17,000-dalton chain = 1.8. This ratio is close enough to 2 to suggest a stoichiometry of 1:1. However, the deviation observed illustrates the fact that all proteins do not bind the Coomassie blue similarly.

Question 7. The exact values of Stokes radii are: ATCase, 6.0 nm; Cat, 4.0 nm; Reg, 2.75 nm. These values are calculated from the molecular masses (Table 1) and from the sedimentation coefficients, which are 11.25S, 5.85S, and 2.8S, respectively. Concerning Reg see answer to Question 5.

Question 8. The R_S/R_{min} values are between 1 and 1.2. Therefore, the proteins studied are globular (Tanford, 1961).

Question 9. The hydrodynamic molecular masses divided by the apparent molecular masses, obtained by SDS–gel electrophoresis, give the number of monomers in the catalytic and regulatory subunits (three and two, respectively).

Question 10. Two kinds of experiments allow us to conclude that native ATCase contains more C than R on a mass basis: the SDS–gel electrophoresis (see answer to Question 6) and the recombination experiment (Chapter 3) performed in the presence of equal amounts of Cat and Reg. When the enzyme reconstituted under these conditions is analyzed by gel electrophoresis under nondenaturing conditions, one can observe an excess of Reg.

The $2(C_3)3(R_2)$ arrangement is the only possible one, since the existence of $1(C_3)6(R_2)$ and $3(C_3)$, whose mass would be identical to that of ATCase, is not compatible with the results discussed above.

Question 11. Reg does not have any catalytic activity. If a catalytic activity is detected, it results from a contamination by the undissociated enzyme E, a phenomenon that is sometime observed when the dissociation reaction was not properly performed. ATP acts as an activator of the reaction, and CTP acts as an inhibitor.

Question 12. At high concentrations of enzyme, substrate is not saturating throughout the incubation. At the end of this incubation, the conditions of initial velocity are no longer met. The decrease in substrate concentration during the test accounts for the decrease in reaction rate and for the fact that the rate of reaction is not proportional to the amount of enzyme. We are dealing here with the concentration of the substrate normalized to its dissociation constant $[S]/K_D$. Under the conditions used here, the substrate is not saturating and the lack of proportionality results from the fact that at high enzyme concentrations the amount of substrate consumed at the end of the test is no longer negligible with respect to the initial amount.

Question 13. The relative molecular masses of catalytic and regulatory **chains** are 33,971 and 16,970, respectively. Thus, in E (M_r = 305,646), the regulatory subunits that do not participate in catalysis represent one-third of the amount of protein. When equal amounts of E and C are used, C will be more active by one-third. Of course, this is true only under the conditions of initial velocity. In the presence of 5-mM aspartate, this difference is even larger as a consequence of the sigmoidicity of the aspartate saturation curve in the case of E.

Question 14. In the presence of very high amounts of enzyme, the reaction equilibrium (plateau) is reached during the incubation.

Question 15. This experiment allows one to determine the amounts of enzyme and time of reaction that ensure measurement of the initial velocities.

Question 16. The saturation curve of C is hyperbolic (Michaelian) and that of E is sigmoidal, showing the existence of cooperativity between the catalytic sites in this enzyme. These differences appear in all of the graphs used.

Question 17. At a given aspartate concentration, ATP increases and CTP decreases the rate of reaction. The sigmoidicity of the saturation curve is increased in the presence of CTP and decreased in the presence of ATP.

Question 18. UTP alone has no effect on the reaction rate. CTP has a limited influence: at high CTP concentration, a plateau is reached. The simultaneous presence of these two effectors leads to a complete inhibition of enzyme activity. It has been observed that the isolated regulatory subunits show negative cooperative effects for CTP binding. One can hypothesize that the binding of this nucleotide to a regulatory site decreases the affinity of the complementary site for CTP but at the same time increases its affinity for UTP. This is currently under investigation.

Question 19. This experiment shows the competitive effects of ATP and CTP. High concentrations of CTP displace ATP entirely from the regulatory site, and only its inhibitory effect is observed. At the point where the curve crosses the abscissa, the effects of the two effectors are balanced on the basis of the respective affinities of CTP and ATP for the regulatory sites and of the quantitative effects that they have on the rate of reaction.

Question 20. The pH dependence for activity of C is independent of the concentration of aspartate. On the other hand, the optimum pH for activity of E is shifted from 6.8 to 8.2 when the concentration of aspartate increases. Between 1- and 20-mM aspartate, E goes from a predominance of the *T* form to a predominance of the *R* form. This correlation suggests that the pK_as of some groups of amino acids located in the catalytic sites are different in the *T* and *R* states of E. C shows the same pH dependence as the *R* form of the native enzyme regardless of the aspartate concentration.

Question 21. ATP increases the rate of the reaction through a mechanism that is distinct from the *T*-to-*R* transition.

Question 22. Native ATCase is more thermostable than the isolated catalytic subunits. However, it slowly dissociates into subunits, and then its inactivation curve parallels that of C.

Question 23. Succinate is a competitive inhibitor with respect to aspartate.

Question 24. C is progressively inhibited by the competitive inhibitor succinate. E is stimulated at low concentrations of succinate. This is due to the fact that at low concentrations of aspartate, where E is essentially under the *T* form, succinate stimulates the activity by inducing the *T*-to-*R* transition. In the presence of higher concentrations, only the competitive inhibition is visible.

References

Ackers, G. K., 1967, A new calibration procedure for gel filtration columns, *J. Biol. Chem.* **242**:3237–3238.

Adair, G. S., 1925, The hemoglobin system. VI. The oxygen dissociation curve of hemoglobin, *J. Biol. Chem.* **63**:529–545.

Allewell, N. M., 1989, *Escherichia coli* aspartate transcarbamoylase: structure, energetics and catalytic and regulatory mechanisms, *Annu. Rev. Biophys. Biophys. Chem.* **18**:71–92.

Batelier, G., 1979, *La Pratique de l'Ultracentrifugation Analytique*, Masson, Paris.

Beaven, G. H., and Holiday, E. R., 1952, Ultraviolet absorption spectra of proteins and amino acids, *Adv. Protein Chem.* **7**:319–386.

Belkaïd, M., Penverne, B., Denis, M., and Hervé, G., 1987, In situ behaviour of the pyrimidine pathway enzymes in *Saccharomyces cerevisiae*. 2. Reaction mechanism of aspartate transcarbamylase dissociated from carbamylphosphate synthetase by genetic alteration, *Arch. Biochem. Biophys.* **254**:568–578.

Bencze, W. L., and Schmid, K., 1957, Determination of tyrosine and tryptophan in proteins, *Anal. Chem.* **29**:1193–1196.

Berger, S. L., and Kimmel, A. R. (eds.), 1987, Guide to molecular cloning techniques, in: *Methods in Enzymology*, vol. 152, Academic Press, San Diego, California.

Brochard, F., Ghazi, M., le Maire, M., and Martin, M., 1989, Size exclusion chromatography on porous fractals, *Chromatographia* **27**:257–263.

Cantor, C. R., and Schimmel, P. R., 1980, *Biophysical Chemistry*, Freeman, San Francisco.

Christopherson, R. I., and Finch, L. R., 1977, Regulation of aspartate carbamoyltransferase of *Escherichia coli* by interrelationship of magnesium and nucleotides, *Biochim. Biophys. Acta* **481**:80–85.

Cohlbert, J. A., Pigiet, V. P., Jr., and Schachman, H. K., 1972, Structure and arrangement of the regulatory subunits in aspartate transcarbamylase, *Biochemistry* **11**:3396–3411.

Cohn, E. J., and Edsall, J. T., 1943, in: *Proteins, Amino Acids and Peptides as Ions and Dipolar Ions* (E. J. Cohn and J. T. Edsall eds.), p. 375, Reinhold, New York.

Collins, K.D., and Stark, G. R., 1971, Aspartate transcarbamylase. Interaction with the transition state analog *N*-(phosphonacetyl)-*L*-aspartate, *J. Biol. Chem.* **246**:6599–6605.

Cox, J. A., and Stein, E. A., 1981, Characterization of a new sarcoplasmic calcium-binding protein with magnesium-induced cooperativity in the binding of calcium, *Biochemistry* **20:**5430–5436.

Dixon, M., 1953, The determination of enzyme inhibitor constants, *Biochem. J.* **55:** 170–171.

Dubin, S. B., Benedek, G. B., Bancroft, F. C., and Freifelder, D., 1970, Molecular weights of coliphages and coliphage DNA. II. Measurement of diffusion coefficients using optical mixing spectroscopy, and measurement of sedimentation coefficients, *J. Mol. Biol.* **54:**547–556.

Eadie, G. S., 1952, On the evaluation of the constants V_m and K_M in enzyme reactions, *Science* **116:**688.

Edsall, J. T., 1970, Definition of molecular weight, *Nature* (London) **228:**888–889.

Eisenberg, H., 1976, *Biological Macromolecules and Polyelectrolytes in Solution*, Clarenton Press, Oxford.

Eisenthal, R., and Cornish-Bowden, A., 1974, The direct linear plot; a new graphical procedure for estimating enzyme kinetic parameters, *Biochem. J.* **139:**715–720.

Eriksson, S., and Sjöberg, B. M., 1989, Ribonucleotide reductase, in: *Allosteric Enzymes* (G. Hervé ed.), chapter 8, CRC Press, Boca Raton, Fla.

Evans, D. R.., Pastra-Landis, S. C., and Lipscomb, W. N., 1975, Isolation and properties of a species produced by the partial dissociation of aspartate transcarbamylase from *Escherichia coli*, *J. Biol. Chem.* **250:**3571–3583.

Fasman, G. D. (ed.), 1976, *Handbook of Biochemistry and Molecular Biology*, CRC Press, Cleveland.

Foote, J., and Schachman, H. K., 1985, Homotropic effects in aspartate transcarbamylase. What happens when the enzyme binds a single molecule of the bisubstrate analog *N*-(phosphonacetyl)-*L*-aspartate, *J. Mol. Biol.* **186:**175–184.

Gerhart, J. C., and Holoubek, H., 1967, The purification of aspartate transcarbamoylase of *Escherichia coli* and separation of its protein subunits, *J. Biol. Chem.* **242:**2886–2892.

Gerhart, J. C., and Pardee, A. B., 1963, The effect of the feedback inhibitor CTP, on subunit interactions in aspartate transcarbamylase, *Cold Spring Harbor Symp. Quant. Biol.* **28:**491–496.

Hanes, C. S., 1932, Studies on plant amylases. I. The effect of starch concentration upon the velocity of hydrolysis by the amylase of germinated barley, *Biochem. J.* **26:** 1406–1421.

Henis, Y., and Levitzki, A., 1989, Mammalian glyceraldehyde-3-phosphate dehydrogenase and its use to elucidate molecular mechanisms of cooperativity, in: *Allosteric Enzymes* (G. Hervé, ed.), Chapter 6, CRC Press, Boca Raton, Fla.

Henri, V., 1903, *Lois Générales de l'Action des Diastases*, Hermann, Paris.

Hervé, G., 1989, Aspartate transcarbamylase from *Escherichia coli*, in: *Allosteric Enzymes* (G. Hervé, ed.), chapter 3, CRC Press, Boca Raton, Fla.

Hervé, G., and Stark, G. R., 1967, Aspartate transcarbamylase. Amino-terminal analyses and peptide maps of the subunits, *Biochemistry* **6:**3743–3747.

Hervé, G., Moody, M. F., Tauc, P., Vachette, P., and Jones, P. T., 1985, Quaternary structure changes in aspartate transcarbamylase studied by X-ray solution scattering; signal transmission following effector binding, *J. Mol. Biol.* **185:**189–199.

Hill, A. V., 1913, The combinations of hemoglobin with oxygen and with carbon monoxide, *Biochem. J.* **7**:471–480.

Hofstee, B. H. J., 1952, On the evaluation of the constants V_m and K_M in enzyme reactions, *Science* **116**:329–331.

Honzatko, R. B., Crawford, J. L., Monaco, H. L., Ladner, J. E., Edwards, B. F. P., Evans, D. R., Warren, S. G., Wiley, D. C., Ladner, R. C., and Lipscomb, W. N., 1982, Crystal and molecular structures of native and CTP-liganded aspartate carbamoyl transferase from *Escherichia coli*, *J. Mol. Biol.* **160**:219–263.

Hoover, T. A., Roof, W. D., Foltermann, K. F., O'Donovan, G. A., Bencini, D. A., and Wild, J. R., 1983, Nucleotide sequence of the structural gene (pyrB) that encodes the catalytic polypeptide of aspartate transcarbamylase of *Escherichia coli*, Proc. Natl. Acad. Sci. USA **80**:2462–2466.

Horiike, K., Tojo, H., Yamano, T., and Nozaki, M., 1983, Interpretation of the Stokes radius of macromolecules determined by gel filtration chromatography, *J. Biochem.* **93**:99–106.

Howlett, G. J., and Schachman, H. K., 1977, Allosteric regulation of aspartate transcarbamylase. Changes in the sedimentation coefficient promoted by the bisubstrate analogue *N*-(phosphonacetyl)-*L*-aspartate, *Biochemistry* **16**:5077–5083.

Hsuanyu, Y., and Wedler, F., 1987, Kinetic mechanism of native *E. coli* aspartate transcarbamylase, *Arch. Biochem. Biophys.* **259**:316–330.

Hsuanyu, Y., and Wedler, F., 1988, Effectors of *E. coli* aspartate transcarbamylase differentially perturb aspartate binding rather than the T–R transition, *J. Biol. Chem.* **263**:4172–4178.

Jaenicke, R., 1987, Folding and association of proteins, *Prog. Biophys. Mol. Biol.* **49**:117–237.

Ke, H., Lipscomb, W. N., Cho, Y., and Honzatko, R. B., 1988, Complex of *N*-phosphonacetyl-*L*-aspartate with aspartate carbamoyltransferase; X-ray refinement, analysis of conformational changes and catalytic and allosteric mechanisms, *J. Mol. Biol.* **204**: 725–747.

Kerbiriou, D., Hervé, G., and Griffin, J. H. 1977, An aspartate transcarbamylase lacking catalytic subunit interactions. Study of conformational changes by ultraviolet absorbance and circular dichroism spectroscopy, *J. Biol. Chem.* **252**:2881–2890.

Konigsberg, W. H., and Henderson, L., 1983, Amino acid sequence of the catalytic subunit of aspartate transcarbamylase from *Escherichia coli*, *Proc. Natl. Acad. Sci. USA* **80**:2467–2471.

Koshland, D. E., Nemethy, G., and Filmer, D., 1966, Comparison of experimental binding data and theoretical models in proteins containing subunits, *Biochemistry* **5**:365–385.

Krause, K. L., Volz, K. W., and Lipscomb, W. N., 1987, The 2.5 Å structure of aspartate carbamoyltransferase complexed with the bisubstrate analogue *N*-(phosphonacetyl)-*L*-aspartate, *J. Mol. Biol.* **193**:527–553.

Ladjimi, M. M., and Kantrowitz, E. R., 1988, A possible model for the concerted allosteric transition as deduced from site-directed mutagenesis studies, *Biochemistry* **27**:276–283.

Laemmli, U. K., 1970, Cleavage of structural proteins during the assembly of the head of bacteriophage T4, *Nature* (London) **227**:680–685.

Laurent, T. C., and Killander, J., 1964, A theory of gel filtration and its experimental verification, *J. Chromatogr.* **14**:317–330.

Léger, D., and Hervé, G., 1988, Allostery and pK_a changes in aspartate transcarbamylase from *Escherichia coli*. I. Analysis of the pH dependence in the isolated catalytic subunits, *Biochemistry* **27**:4293–4298.

Lehninger, A. L., 1975, *Biochemistry*, Worth Publishers, New York.

le Maire, M., 1987, Utilisation de la chromatographie sur gel pour la détermination de la taille et de la masse moléculaire des protéines, *Biosciences* **6**:119–123.

le Maire, M., Lind, K. E., Jorgensen, K. E., Røigaard, H., and Møller, J. V., 1978, Enzymatically active Ca^{2+}-ATPase from salrcoplasmic reticulum membranes, solubilized by nonionic detergents: role of lipids for aggregation of the protein, *J. Biol. Chem.* **253**:7051–7060.

le Maire, M., Rivas, E., and Møller, J. V., 1980, Use of gel chromatography for determination of size and molecular weight of proteins: further caution, *Anal. Biochem.* **106**:12–21.

le Maire, M., Møller, J. V., and Tardieu, A., 1981, Shape and thermodynamic parameters of a Ca^{2+}-dependent ATPase. A solution X-ray scattering and sedimentation equilibrium study, *J. Mol. Biol.* **150**:273–296.

le Maire, M., Aggerbeck, L. P., Monteilhet, C., Andersen, J. P., and Møller, J. V., 1986, The use of high-performance liquid chromatography for the determination of size and molecular weight of proteins: a caution and a list of membrane proteins suitable as standards, *Anal. Biochem.* **154**:525–535.

le Maire, M., Ghazi, A., Møller, J. V., and Aggerbeck, L. P., 1987, The use of gel chromatography for the determination of sizes and relative molecular masses of proteins. Interpretation of calibration curves in terms of gel-pore-size distribution, *Biochem J.* **243**:399–404.

le Maire, M., Viel, A., and Møller, J. V., 1989a, Size exclusion chromatography and universal calibration of gel columns, *Anal. Biochem.* **177**:50–56.

le Maire, M., Ghazi, A., Martin, M., and Brochard, F., 1989b, Calibration curves for size-exclusion chromatography: description of HPLC gels in terms of porous fractals, *J. Biochem.* **106**:814–817.

Lineweaver, H., and Burk, D., 1934, The determination of enzyme dissociation constant, *J. Am. Chem. Soc.* **56**:658–666.

Lowry, O. H., Rosebrough, N. J., Farr, A. L. and Randall, R. J., 1951, Protein measurement with Folin phenol reagent, *J. Biol. Chem.* **193**:265–275.

Luzzati, V., and Tardieu, A., 1980, Recent developments in solution X-ray scattering studies of structures of biological interest, *Annu. Rev. Biophys. Bioeng.* **9**:1–29.

Makino, S., 1979, Interaction of proteins with amphiphilic substances, *Adv. Biophys.* **12**:131–184.

Matsudaira, P., 1987, Sequence from picomole quantities of proteins electroblotted onto polyvinylidene difluoride membranes, *J. Biol. Chem.* **262**:10035–10038.

McEwen, C. R., 1967, Tables for estimating sedimentation through linear concentration gradients of sucrose solution, *Anal. Biochem.* **20**:114–149.

Merril, C. R., Goldman, D., Sedman, S. A., and Ebert, M. H., 1981, Ultrasensitive stain for proteins in polyacrylamide gels shows regional variation in cerebrospinal fluid proteins, *Science* **211**:1437–1438.

Metzler, D. E., Harris, C., Yang, I. Y., Siano, D., and Thomson, J. A., 1972, Band-shape analysis and display of fine structure in protein spectra: a new approach to perturbation spectroscopy, *Biochem. Biophys. Res. Commun.* **46**:1588–1597.

Michaelis, L., and Menten, M. L., 1913, Die Kinetik der Invertinwirkung, *Biochem. Z.* **49**:333–369.

Minton, A. P., 1989, Analytical centrifugation with preparative ultracentrifuges, *Anal. Biochem.* **176**:209–216.

Møller, J. V., le Maire, M., and Andersen, J. P., 1986, in: *Progress in Protein–Lipid Interactions* (A. Watts and J. J. H. H. M. DePont, eds.), vol. 2, chapter 5, pp. 147–196, Elsevier, Amsterdam/New York.

Monod, J., Changeux, J. P., and Jacob, F., 1963, Allosteric proteins and cellular control systems, *J. Mol. Biol.* **6**:306–329.

Monod, J., Wyman, J., and Changeux, J. P., 1965, On the nature of allosteric transitions: a plausible model, *J. Mol. Biol.* **12**:88–118.

Nomenclature Committee of the International Union of Biochemistry (NC-IUB), 1979, Units of enzyme activity. Recommendations 1978, *Eur. J. Biochem.* **97**:319–320.

Parsegian, V. A., Rand, R. P., Fuller, N. L., and Rau, D. C., 1986, Osmotic stress for the direct measurement of intermolecular forces, *Methods Enzymol.* **127**:400–416.

Perbal, B., and Hervé, G., 1972, Biosynthesis of *Escherichia coli* aspartate transcarbamylase. I. Parameters of gene expression and sequential biosynthesis of the subunits, *J. Mol. Biol.* **70**:511–529.

Perbal, B., Gueguen, P., and Hervé, G., 1977, Biosynthesis of *Escherichia coli* aspartate transcarbamylase. II. Correlated biosynthesis of the catalytic and regulatory chains and cytoplasmic association of the subunits, *J. Mol. Biol.* **110**:319–340.

Pollet, R. J., Haase, B. A., and Standaert, M., 1979, Macromolecular characterization by sedimentation equilibrium in the preparative ultracentrifuge, *J. Biol. Chem.* **254**:30–33.

Porath, J., 1963, Some recently developed fractionation procedures and their application to peptide and protein hormones, *Pure Appl. Chem.* **6**:233–244.

Porter, R. W., Modebe, M. O., and Stark, G. R., 1969, Aspartate transcarbamylase: kinetic studies of the catalytic subunit, *J. Biol. Chem.* **244**:1846–1859.

Prescott, L. M., and Jones, M. E., 1969, Modified methods for the determination of carbamyl aspartate, *Anal. Biochem.* **32**:408–419.

Reynolds, J. A., and Tanford, C., 1970, The gross conformation of protein-sodium dodecyl sulfate complexes, *J. Biol. Chem.* **245**:5161–5165.

Ricard, J., 1989, Concepts and models of enzyme cooperativity, in: *Allosteric Enzymes* (G. Hervé, ed.), Chapter 1, CRC Press, Boca Raton, Fla.

Rizzolo, L. J., le Maire, M., Reynolds, J. A., and Tanford, C., 1976, Molecular weights and hydrophobicity of the polypeptide chain of sarcoplasmic reticulum calcium (II) adenosine triphosphatase and of its primary tryptic fragments, *Biochemistry* **15**:3433–3437.

Roland, K. L., Powell, F. E., and Turnbough, C. L., 1985, Role of translation and attenuation in the control of pyrBI operon expression in *Escherichia coli* K–12, *J. Bacteriol.* **163**:991–999.

Roof, W. D., Foltermann, K. F., and Wild, J. R., 1982, The organization and regulation of the pyrBI operon in *E. coli* includes a Rho-independent attenuator sequence, *Mol. Gen. Genet.* **187**:391–400.

Royer, C. A., Tauc, P., Hervé, G., and Brochon, J. C., 1987, Ligand binding and protein dynamics: a fluorescence depolarization study of aspartate transcarbamylase from *Escherichia coli*, *Biochemistry* **26**:6472–6478.

Schachman, H. K., Pauza, C. D., Navre, M., Karels, M. J., Wu, L., and Yang, Y. R., 1984, Location of amino acid alterations in mutants of aspartate transcarbamoylase: structural aspects of interallelic complementation, *Proc. Natl. Acad. Sci. USA* **81**:115–119.

Siegel, L. M., and Monty, K. J., 1966, Determination of molecular weights and frictional ratios of proteins in impure system by use of gel filtration and density gradient centrifugation. Application to crude preparations of sulfite and hydroxylamine reductases, *Biochim. Biophys. Acta* **112**:346–362.

Smith, C. A., 1988, Estimation of sedimentation coefficients and frictional ratios of globular proteins, *Biochem. Educ.* **16**:104–106.

Steele, J. C., Tanford, C., and Reynolds, J. A., 1978, Determination of partial specific volumes for lipid-associated proteins, *Methods Enzymol.* **48F**:11–23.

Subramani, S., and Schachman, H. K., 1980, Mechanism of disproportionation of aspartate transcarbamylase molecules lacking one regulatory subunit, *J. Biol. Chem.* **255**:8136–8144.

Tanford, C., 1961, *Physical Chemistry of Macromolecules*, John Wiley and Sons, New York.

Tanford, C., Nozaki, Y., Reynolds, J. A., and Makino, S., 1974, Molecular characteristics of proteins in detergent solutions, *Biochemistry* **13**:2369–2376.

Tardieu, A., Vachette, P., Gulik, A., and le Maire, M., 1981, Biological macromolecules in solvent of variable density: characterization by sedimentation equilibrium, densimetry and X-ray forward scattering and an application to the 50S ribosomal subunit from *Escherichia coli*, *Biochemistry* **20**:4399–4406.

Tauc, P., Leconte, C., Kerbiriou, D., Thiry, L., and Hervé, G., 1982, Coupling of homotropic and heterotropic interactions in *Escherichia coli* aspartate transcarbamylase, *J. Mol. Biol.* **155**:155–168.

Thiry, L., and Hervé, G., 1978, The stimulation of *Escherichia coli* aspartate transcarbamylase activity by adenosine triphosphate. Relations with the other regulatory conformational changes: a model, *J. Mol. Biol.* **125**:515–534.

Towbin, H., Staehelin, T., and Gordon, J., 1979, Electrophoretic transfer of proteins from polyacrylamide gels to nitrocellulose sheets: procedure and some applications, *Proc. Natl. Acad. Sci. USA* **76**:4350–4354.

Vérétout, F., Delaye, M., and Tardieu, A., 1989, Molecular basis of eye lens transparency. Osmotic pressure and X-ray analysis of alpha-crystallin solutions, *J. Mol. Biol.* **205**:713–728.

Viratelle, O., and Seydoux, F. J., 1975, Pseudo-conservative transition: a two state model for the cooperative behavior of oligomeric proteins, *J. Mol. Biol.* **92**:193–205.

Volz, K. W., Krauze, K. L., and Lipscomb, W. N., 1986, The binding of *N*-(phosphonacetyl)-*L*-aspartate to aspartate transcarbamyoltransferase of *Escherichia coli*, *Biochem. Biophys. Res. Commun.* **136**:822–826.

Weber, G., 1972, Ligand binding and internal equilibria in proteins, *Biochemistry* **11**:864–878.

Weber, K., 1968, New structural model of *E. coli* aspartate transcarbamylase and the amino-acid sequence of the regulatory polypeptide chain, *Nature* (London) **218**:1116–1119.

Weber, K., and Osborn, M., 1969, The reliability of molecular weight determinations by dodecyl sulfate-polyacrylamide gel electrophoresis, *J. Biol. Chem.* **244**:4406–4412.

Whitaker, J. R., and Granum, P. E., 1980, An absolute method for protein determination based on difference in absorbance at 235 and 280 nm, *Anal. Biochem.* **109**:156–159.

Wild, J. R., Loughrey-Chen, S. J., and Corder, T. S., 1989, In the presence of CTP, UTP becomes an allosteric inhibitor of aspartate transcarbamylase, *Proc. Natl. Acad. Sci.* **86**:46–50.

Zaccai, G., and Jacrot, B., 1983, Small angle neutron scattering, *Annu. Rev. Biophys. Bioeng.* **12**:139–157.

Index

Printed in the United States
by Baker & Taylor Publisher Services

Printed in the United States
by Baker & Taylor Publisher Services